STUDENT STUDY GUIDE
to accompany

FUNDAMENTALS OF CHEMISTRY

Second Edition

David E. Goldberg
Brooklyn College

Prepared by
Margaret G. Kimble
Indiana University–Purdue University

Boston, Massachusetts Burr Ridge, Illinois Dubuque, Iowa
Madison, Wisconsin New York, New York San Francisco, California St. Louis, Missouri

WCB/McGraw-Hill

A Division of The McGraw·Hill Companies

Student Study Guide to accompany
FUNDAMENTALS OF CHEMISTRY

Copyright ©1998 by The McGraw-Hill Companies, Inc. All rights reserved. Printed in the United States of America. The contents of, or parts thereof, may be reproduced for use with
FUNDAMENTALS OF CHEMISTRY
Second Edition
David E. Goldberg
provided such reproductions bear copyright notice and may not be reproduced in
any form for any other purpose without permission of the publisher.

Acid free paper
This book is printed on acid-free paper .

1 2 3 4 5 7 8 9 0 QPD/QPD 9 0 9 8 7

ISBN 0-697-29152-9

www.mhhe.com

This book is dedicated to my loving husband, Edward, who has endured many long hours of my absence and, more importantly, to God, who loves all of us all of the time.

Table of Contents

Introduction	How to Study Chemistry	ix
Chapter 1	**Basic Concepts**	**1**
	Sample Exam Questions	9
	Toothpaste Boat	13
	Self-Inflating Balloons	13
Chapter 2	**Measurement**	**15**
	Sample Exam Questions	23
	Introduction to Thermodynamics	26
Chapter 3	**Atoms and Atomic Masses**	**27**
	Sample Exam Questions	33
	Visualizing Emission Spectra	37
Chapter 4	**Electronic Configuration of the Atom**	**39**
	Sample Exam Questions	46
	Fun with Fireworks	50
Chapter 5	**Chemical Bonding**	**51**
	Sample Exam Questions	57
	Growing Crystals	61
Chapter 6	**Nomenclature**	**63**
	Sample Exam Questions	67
	Household Nomenclature	72
Chapter 7	**Formula Calculations**	**73**
	Sample Exam Questions	79
	Fun with Food	83
Chapter 8	**Chemical Reactions**	**85**
	Sample Exam Questions	91
	Acid and Base Test Solutions	95
Chapter 9	**Net Ionic Equations**	**97**
	Sample Exam Questions	101
	Chromatography	104
Chapter 10	**Stoichiometry**	**105**
	Sample Exam Questions	110
	Baking a Cake	114
Chapter 11	**Molarity**	**115**
	Sample Exam Questions	119
Chapter 12	**Gases**	**123**
	Sample Exam Questions	131
	The Power of Air	135
Chapter 13	**Atomic and Molecular Properties**	**137**
	Sample Exam Questions	143

Chapter 14	**Solids and Liquids**	**145**
	Sample Exam Questions	151
Chapter 15	**Solutions**	**153**
	Sample Exam Questions	159
	Inside-Out Bubbles	162
Chapter 16	**Oxidation Numbers**	**165**
	Sample Exam Questions	171
	Concentration of Oxygen in Air	174
Chapter 17	**Reaction Rates and Equilibrium**	**177**
	Sample Exam Questions	181
	Photochromic and Thermochromic Reactions	185
Chapter 18	**Acids and Bases**	**187**
	Sample Exam Questions	193
	Red Cabbage Indicator Solution	197
Chapter 19	**Organic Chemistry**	**199**
	Sample Exam Questions	208
	Grandma's Lye Soap	212
Chapter 20	**Nuclear Chemistry**	**213**
	Sample Exam Questions	219

Answers to Sample Exam Questions   223

Introduction

How to Study Chemistry

Many students apprehensively approach their first chemistry course and postpone it as long as possible. But, if you can give chemistry a chance, it will prove to be an interesting and exciting subject. Do not carry any prejudices into the beginning of this course, but instead look forward to learning about the marvelous world around you in a way you have never done before.

After assembling your textbooks, notes, paper, pens and calculator, you must spend some time deciding when, where and how fast you will study. These decisions must be carefully individualized to suit your own abilities and circumstances.

The **time** for study will vary from one student to another, but as a rule you should set aside at least two hours for each credit hour you are taking. If this means needing additional baby-sitting or time off from work, understand that this must be considered when scheduling a college class. Time of day also plays an important role. Some people learn better "at the crack of dawn." Other people do best when they are "burning the midnight oil." You should know by now which type of person you are and should plan your study sessions accordingly. The important thing is that you be consistent. Study at the same time every day and do it with the same commitment you would have if you were going to class.

By carefully selecting your study **place** you can greatly increase your chances of good performance in this course. You must have a quiet place with good lighting, comfortable conditions, and, most importantly, freedom from interruptions. You may need to find a place to study that is neither at school or at home. A quiet library room, a study area at a local church, or a study table at your fraternity or sorority house could all be possibilities. Each student will have individual requirements for what is necessary to achieve concentration and learning.

In addition to establishing the time and place to study, the successful student will also carefully plan the **pace** at which the material can be mastered. Chemistry is a subject that is best learned by working with it every day. If you have ever studied a musical instrument or learned to type, then you have experienced the consistency of effort that is required of chemistry students. You must "practice" every day in order to learn the material. You cannot wait until the day before the exam to start to cram the information. You must work diligently and learn the methods used in solving problems that are used in chemistry.

It is highly recommended that you establish a study group during the first week of classes. This serves several purposes. Primarily, it gives you a camaraderie at a time when you may really need a friend. It makes you accountable for study habits. And, a group provides a source of lecture notes in the event of an unexpected absence from class.

There are some easy ways to enhance your learning. Flash cards are useful. Write out a sample problem for each type of calculation you learn on one side of each card. On the other side write a detailed solution of that problem. You might even include text references where that material can be found for review. As you complete a chapter, shuffle the cards with ones from previous chapters. This will provide an excellent review for a comprehensive test.

It is important to realize that your efficiency of learning is not 100%. When you sit down to study, remember there is a world of things determined to interrupt you. The time you spend sitting at a desk with your book open will not necessarily reflect the time you have spent in actual learning. You must figure out approximately what your efficiency is and adjust your time so that you can accommodate the interruptions. There are two things that you might try to increase your efficiency. First, **be sure to take breaks**. Most college classes are 50 minutes long. This is the attention span of an average adult. By arranging a ten-minute break between 50-minute sessions you can optimize your ability to concentrate. Be careful though, those distractions are out there to get you and if they see you "killing time," they will think you are available for other activities. Return to the studying just as if you were returning to a scheduled class. Second, **be sure you are not tired**. Always start studying with the subject that is the hardest to comprehend. When you tire, you can study one that is more easily understood. Also, after some trial and error, you will find the time of day when you and chemistry can get along best.

Integrating the textbook with the lecture is indeed an art. It is very important in a science course to have read the material before coming to the lecture about that material. You should have in mind the objectives of each section as well as the terminology that may be new to you. Knowing what is in the text will save much time when trying to write legible lecture notes. By knowing what is in the textbook, you will not have to take as many notes. This will give you more time to listen to the lecture.

You must also allocate time for reviewing the lecture notes soon after class. H. F. Spitzer reports in the *Journal of Educational Psychology* that a student who waits 24 hours before reviewing lecture notes will forget 46% of the material presented. In two days, 50% is gone and, at the end of a week 62% is gone. On the other hand, the student who reviews lecture notes within a few hours retains about 98% of what was said, holds 97% a week later and still remembers more than 90% of the lecture three weeks later. This percentage can easily be translated into your grade for the course.

Make yourself an active participant in the class. It is said that the best way to learn a subject is to teach it. To benefit from this, it is suggested that you should try the following:

1) Pretend that you are going to give the lecture on each chapter. Plan what you are going to say. Find extra material in the library to include in the class. Think of some demonstrations that would be of interest.

2) At the end of each chapter write your own exam on the material. Be sure to include extra hard questions over the areas you know the least. Try to integrate material from previous chapters into the new material. Be sure your questions are thoroughly researched for accuracy.

How To Use the Study Guide

This study guide is designed to help you efficiently study the material in each chapter of your textbook. The chapters and sections in the study guide correspond directly to the chapters and sections of your textbook. Whenever there are directions for completing a task, the text is marked with a small box. ☐ These can be checked to indicate your progress through the chapter. The margins of each page allow room for your comments and thoughts as you read each chapter.

Many students find that preparing an outline of the text chapter before going to lecture helps them organize the lecture notes for later study. Be sure to check with your course schedule to find exactly what topics will be covered and read and outline those sections before class.

Begin each chapter by reading the section entitled "Before You Get Started." This will tell you what you need to remember from previous chapters as well as any math skills necessary to work the problems of this chapter. Then follow the directions for working through the chapter. A review section summarizes the material and gives you a chance to make sure that you have learned each important point of the chapter. To check your mastery, do the sample exam questions which follow each chapter.

Since chemistry should be fun, an experiment that you can perform at home is at the end of most chapters. These are written to allow you to have fun outside chemistry class using the information you have learned and supplies you would usually have on hand.

The most important factor in your success as you study this material is you. Spend the time needed to master the material when it is presented. Discipline yourself to study regularly. And, before you know it, you will be on your way to getting that "A" in chemistry!

Chapter 1: Basic Concepts

In this chapter you will be learning some terminology associated with chemistry. In each chapter it is important to see what you are going to learn before you actually start learning it. Open your book to Chapter 1 and read quickly through the pages. If something catches your eye, spend some time reading that part more carefully. Be sure to look at the pictures and read the captions. Read the objective that is found at the beginning of each section. Look at the tables and charts. Know where to find them and what is in each so you can refer to them as you read the chapter.

Section 1.1: Classification of Matter

☐ Read the section before continuing. There are several words introduced in this section which you will use later in the text. Write out definitions for each of these, referring to the text as well as any lecture notes you may have taken on this material. Try to use your own words to define each of them.

chemical change (chemical reaction)	_____
physical change	_____
compound	_____
mixture	_____
heterogeneous	_____
homogeneous	_____

It is the object of a science course to not only teach you the science but to also stimulate your curiosity. Spend some time observing the world around you and see if you can use these words to describe that world to others.

☐ Make sure to read the Examples and work the Practice Problems in the text as you read this section. Check your answers carefully.

☐ Work the Problems for this section at the end of the chapter.

Section 1.2: Properties

In order to describe chemical reactions you will need to be able to recognize the various properties of matter.

☐ Read Section 1.2 and note the words in bold print. As you did in Section 1.1, write definitions for the following:

extensive properties: _____

intensive properties: _____

☐ Before continuing with this chapter, take a moment and look at the photographs. The author has used many excellent illustrations and photographs throughout the text. Use these to help you understand what is being explained. Hopefully you will have the opportunity to apply the illustrated procedures in the laboratory part of your course. If your course does not include a laboratory, you should spend extra time looking at these illustrations so that you can relate your new knowledge of chemistry to the world around you.

☐ Make sure to read the Examples and work the Practice Problems in the text as you read this section. Check your answers carefully.

☐ Work the Problems for this section at the end of the chapter.

Section 1.3: Matter and Energy

Section 1.3 introduces the relationship between matter and energy. Chemistry is the study of matter.

☐ Matter is anything that has _____ and _____ _____.

☐ Name three things that are classified as matter.

     1. _____   2. _____   3. _____

☐ Explain in your own words how mass and weight are different and when they can be considered the same.

☐ Look at Table 1.3. These are examples of energy. Can you give examples of how energy is converted from one form to another in addition to those which follow?

     heat   wood burns and gives off heat
     _____
     _____

mechanical	explosions produce gases that increase pressure and move pistons in a car engine
	_____
	_____
electrical	batteries run radios
	_____
	_____
sound	firecrackers make noise
	_____
	_____
chemical	your body uses food for energy
	_____
	_____
light	wood burns and gives off light
	_____
	_____

☐ The interconversion between matter and energy played an important role in the development of the atomic bomb as explained in the Enrichment topic of Chapter 1. Science fiction writers try to make their stories believable by using good science when they design the machines and processes used in their stories. The works of Gene Roddenberry demonstrate the attention to detail used by good writers. The Star Trek series has even inspired some real science. Spend some time in the next few days thinking about the relationship between matter and energy and the examples above.

☐ Make sure to read the Examples and work the Practice Problems in the text as you read this section. Check your answers carefully.

☐ Work the Problems for this section at the end of the chapter.

Section 1.4: Chemical Symbols

☐ Section 1.4 introduces you to the symbols of the elements. In order to work problems in later chapters you must know the more common symbols. Your instructor may assign certain elements for you to know. Find someplace in your daily routine where you can post these and look at them often. Many students find that memorization comes easiest if it is done at the same time and place every day. If you have a break at work, this could be used for memorization work. Flash cards seem to work best to memorize this type of material. You can carry them with you and refer to them whenever you are waiting somewhere. On your cards you should include the names and symbols of the first 36 elements plus some of the more common ones lower in the periodic table. Be very careful in copying the symbols. If there is just one letter, write only one capital letter. If there are two letters in the symbol -- a capital and a lower case letter, be sure to copy them just as they are written. Table 1.4 lists some of the symbols which do not have the same initials as their names. Some elements originally had Latin or German names. If you know the original name and some of the root word usage in English it is easier to remember these symbols. You do not need to memorize the foreign names.

Gold	Au	Aurum is the Latin word for shining dawn. *Aurora borealis* is the Northern Lights. Remember that gold is shiny and that it has an aura.
Iron	Fe	Ferrum is the Latin word for iron. This metal is mentioned in the Bible. Iron is an essential mineral in nutrition and is supplemented in the form of ferrous sulfate.
Lead	Pb	Plumbum is the Latin word for lead. Carpenters use a "plumb" line, a string with a lead weight on it, to determine true vertical. We call lead pipe experts "plumbers."
Mercury	Hg	Hydrargyrum is the Latin word for "liquid silver." Originally thought to be the liquid state of silver by the ancients, mercury is mentioned in ancient Chinese and Hindu literature.
Potassium	K	Kalium is the Latin word for potassium. Do not confuse the word kalium with calcium.
Silver	Ag	Argentum is the Latin word for silver.
Sodium	Na	Natrium is the Latin word for sodium. A natatorium is a swimming area. The ocean is the most common source of sodium because of the salt, sodium chloride. Since ancient Romans used the oceans as their source of salt as well as a place to swim, we have a connection between the two words.
Tin	Sn	Stannum is the Latin word for tin. In Britain, a stannary is a place where tin is mined or smelted. Stannous fluoride is a compound of tin and fluorine that is applied to children's teeth to prevent decay.

Tungsten   **W**   Wolfram is derived from the German words *wolf* (wolf) and *rahm* (to eat) because it was believed that the ore interfered with the smelting of tin by actually eating the tin. Tungsten is commonly found in electric light filaments.

More about the names of the chemical elements and their history and development can be found in the Chemical Rubber Company *Handbook of Physics and Chemistry*. You can also find information in an unabridged dictionary.

☐ Make sure to read the Examples and work the Practice Problems in the text as you read this section. Check your answers carefully.

☐ Work the Problems for this section at the end of the chapter.

Section 1.5: The Periodic Table

Section 1.5 introduces you to the most important table in the book: the periodic table. You will be learning many relationships between elements and predicting many properties of elements based on their location in the periodic table. There are several blank periodic tables throughout this study guide which you may use to study the material. As you learn about the parts of the table you can organize your notes. Most instructors provide a periodic table with exams so it becomes very handy to learn your chemistry while looking at a periodic table.

☐ Where would you find each of the following in the periodic table? Give the names and symbols of several members of each classification. Locate these on the blank periodic table on the next page. Be sure to keep this for review before your first exam.

a period	
a group	
alkali metal	
alkaline earth metal	
halogen	
noble gas	
coinage metal	

main group elements	
transition elements	
inner transition elements	
metals	
nonmetals	
metalloids	
lanthanides	
actinides	

Give several differences between metals and nonmetals. (See textbook)

Metals   **Nonmetals**

_____   _____

_____   _____

_____

☐ Make sure to read the Examples and work the Practice Problems in the text as you read this section. Check your answers carefully.

☐ Work the Problems for this section at the end of the chapter.

Section 1.6: Laws, Hypotheses, and Theories

Section 1.6 tells you about the development process of scientific thought. Scientists communicate with each other on a regular basis through journals and meetings. When significant findings are made, the whole community of scientists is informed. For this reason, scientists are trying to standardize the methods of measurement so that other scientists in all parts of the world can learn from the same observations. In Chapter 2 you will learn about these measurements.

☐ In your own words, explain what a hypothesis is.

_____

_____

☐ What is a theory?

_____

_____

☐ What is a law?

_____

_____

☐ Make sure to read the Examples and work the Practice Problems in the text as you read this section. Check your answers carefully.

☐ Work the Problems for this section at the end of the chapter.

Chapter 1: Finishing Up

☐ Carefully read the summary section at the end of the chapter. Do you understand each paragraph? Do you know the terms used? If not, review the section indicated at the end of the paragraph.

☐ The section titled "Items for Special Attention" gives you some pointers to help you learn the material correctly and also can keep you from making some of the more common mistakes. Read these carefully.

☐ Work the "Self-Tutorial Problems." Check each answer as you complete the problem. If you have trouble with one of these, be sure to ask your instructor for help.

☐ Look over the list of "Key Terms" at the beginning of the chapter. If you do not recognize a term or are unsure of how it was used in the chapter, go back to that section and reread it.

☐ Use your lecture notes and the text to find the important topics covered in class. Make flash cards so that you can study these carefully before the exam.

☐ Work each of the problems at the end of the chapter in your text. Be sure to check your answers. If the answer is not given for that problem, work another that is similar and check the answer. If you are incorrect, check your work and review the chapter to see if you can answer the question correctly. If you cannot, get help from your instructor.

☐ Work the sample exam questions on the next few pages.

☐ Check your answers when you have completed all of them.

☐ Make up your own exam and exchange it with a fellow student or use it in your study group.

Chapter 1: Sample Exam Questions

1. Which of the following are compounds? Circle your answer(s).

   CO   Na H_2O Cs   No   NO NH_3 Co

2. Which of the following are elements? Circle your answer(s).

   CO   K   HCl $CaCl_2$ Mo   Mn   KCl   Bi

3. Tell what type of change--chemical (c) or physical (p)--is probably occurring when:

   _____ a. a solid is placed in a liquid and bubbles appear

   _____ b. a liquid is heated and the liquid disappears

   _____ c. a solution is allowed to cool and a solid appears in the liquid

   _____ d. two liquids are mixed and a gas is given off

   _____ e. a beaker falls to the floor and breaks

   _____ f. the liquid in a container disappears when it is left on the lab bench overnight

4. Write the names for the following chemical elements:

   (a) Fe _____   (e) Pb _____

   (b) Au _____   (f) Na _____

   (c) Hg _____   (g) Ag _____

   (d) Cu _____   (h) P _____

5. Identify each of the following as either an element (e) or a compound (c):

   (a) CO   _____   (d) No   _____

   (b) NaCl _____   (e) $AlPO_4$ _____

   (c) Co   _____   (f) NO   _____

6. Write the symbol for each of the following:

    (a) Iron _____ (d) Sulfur _____

    (b) Sodium _____ (e) Potassium _____

    (c) Chlorine _____ (f) Gold _____

7. Locate the following on the periodic table below:

    a. halogens   b. noble gases   c. nonmetals   d. metals

8. For each of the following properties, determine whether the property is intensive (i) or extensive (e):

    (a) _____ A substance is yellow.

    (b) _____ A solid sample weighs 5.98 grams.

    (c) _____ Gold is a shiny metal.

    (d) _____ Sugar tastes sweet.

    (e) _____ A sample of liquid has a volume of 12 mL.

9. Identify the following as a chemical (c) or a physical (p) property:

   (a) ____ color

   (b) ____ flammability

   (c) ____ gas at room temperature

   (d) ____ melting point

   (e) ____ biodegradability

10. For each of the following identify whether a chemical (c) or physical (p) change has occurred:

   (a) _____ A sample of metal is heated in air. A white powder is formed that weighs more than the original metal sample.

   (b) _____ A substance is heated. The resulting white powder weighs less.

   (c) _____ A liquid is heated and it evaporates.

   (d) _____ Ice cubes become smaller in the freezer.

11. What is the difference between mass and weight?

12. Tell whether each of the following elements is a Metal (m) or a Nonmetal (n).

   (a) ____ iodine

   (b) ____ barium

   (c) ____ cobalt

   (d) ____ cadmium

   (e) ____ platinum

13. Tell whether each of the following elements is a main group (mg), transition (t), or inner transition (it).

(a) _____ cesium

(b) _____ carbon

(c) _____ iron

(d) _____ uranium

(e) _____ magnesium

14. Tell whether each of the following elements is an alkali metal (am), alkaline earth metal (aem), coinage metal (cm), halogen (h), or noble gas (ng).

(a) _____ potassium

(b) _____ silver

(c) _____ bromine

(d) _____ argon

(e) _____ calcium

15. What is the difference between a *group* and a *period* in the periodic table?

Chapter 1: Experiment at Home

The Toothpaste Boat

Find a small piece of wood. Balsa is best but a piece of bark from a tree will also work well. Place a ribbon of toothpaste about 1/4 inch long across the "stern" (back) of your boat. Launch your boat in a large bowl or in the bathtub. Allow the toothpaste to touch the water. As the toothpaste dissolves in the water, the molecules will be leaving the back end of the boat. This will cause a forward movement to the boat. This is a molecular example of the physical law stating that for every action there is an opposite and equal reaction.

The same thing happens when you jump off a skateboard. If you jump backward, the skateboard is propelled forward.

Self-Inflating Balloons

Place three tablespoons of baking soda in an ordinary balloon. Add three tablespoons of vinegar and tie the balloon quickly. Allow it to stand and it will inflate. The gas formed in the reaction of the baking soda with the vinegar is carbon dioxide. As the gas molecules are formed, they collide with the inside of the balloon building up pressure and causing the balloon to inflate. As more gas molecules are produced, more pressure is exerted and the balloon becomes larger.

Chapter 2: Measurement

Before you get started

Be sure you have a calculator that will do the necessary calculations. These are identified as "scientific" calculators. You should familiarize yourself with the calculator you have selected. You should be able to perform the following:

- Add, subtract, multiply and divide numbers using your calculator
- Know how to enter negative numbers
- Understand the order of operations for calculations
- Multiply fractions
- Enter numbers in exponential notation
- Perform calculations using numbers written in exponential notation

☐ You should study Appendix 1 in the back of your text at this time.

Section 2.1: Factor Label Method

☐ Begin your study of Chapter 2 by reading Section 2.1. Be sure to work <u>all</u> of the Practice Problems in the text. These will help you develop your thinking pattern for handling the more difficult problems later. Some of these may seem very simple, but be sure to set them up as explained in the text.

Do you see a pattern developing? The unit of the quantity you start with is always cancelled when the next factor is used. The numerator of the factor contains the new unit for the measurement. Look carefully for patterns when you are working problems. This will help you establish good learning skills.

☐ Make sure to read the Examples and work the Practice Problems in the text as you read this section. Check your answers carefully.

☐ Work the Problems for this section at the end of the chapter.

Section 2.2: The Metric System

Most of science uses the metric system for measurement. There are several very good reasons for this. First of all, scientists need a common system in which to report data to one another. They need to be able to communicate results and check each others' theories. Secondly, the metric system is versatile. There are units to fit every size of measurement. Finally, the metric system is easy to use. For example, going from one unit of mass to another unit of mass is as simple as moving a decimal point.

Be sure you know what is expected of you for memorization of the conversion tables. Memorizing at least one conversion between English and metric for each of the base units will enable you to make conversions without having to look up the values each time. You may have to use more factors in the conversion, but with the use of your calculator even this can be done quickly. The usual choices are:

LENGTH	1 inch = 2.54 centimeters
MASS	1 pound = 454 grams
VOLUME	1.06 quart = 1 liter

Most Americans are taught the English system in school. We see the English system everywhere. It is important to be able to make approximations between the two systems quickly. If you can visualize the size of some of the metric units you will be able to function very well using metrics.

The unit of **LENGTH**, the **METER**, can be approximated by the length of a yard. Although a yard is slightly shorter than a meter, it still serves as a good device for estimating meters. Another useful approximation is that 80 kilometers is approximately 50 miles. Check out the speedometer in a car. The speed of 50 miles per hour should be about the same as 80 kilometers per hour if the speedometer is calibrated in both units.

The unit of **MASS**, the **GRAM**, is approximately the mass of a paper clip. Much of the English system of weights uses the pound. A useful conversion would be that 1 kilogram is a little more than 2 pounds. This is very useful for those who are trying to diet. Just divide your weight in pounds by two and you will have an instant diet and your approximate mass in kilograms.

The unit of **VOLUME**, the **LITER**, is probably more familiar to you than the others. Since soft drinks now come in 2-liter bottles, you have seen this measure and can easily visualize that a liter is just slightly bigger than a quart, and that a 2-liter bottle is just a little larger than a half gallon.

☐ Make sure to read the Examples and work the Practice Problems in the text as you read this section. Check your answers carefully.

☐ Work the Problems for this section at the end of the chapter.

Section 2.3: Significant Digits

Significant digits are used in science for reporting measured values. Any time a measurement is made it is very important that the reading is reported to the proper number of digits. To determine the number of digits to report, examine the instrument and determine the calibration of the instrument. Is it calibrated in milliliters? Tenths of milliliters? The calibration marks tell you the last certain digit. You must then make a guess at the next digit and report that guess as the estimated digit. When someone else reads your data from the experiment they will know what calibration was on the device you used to make the measurement.

☐ Read Section 2.3.

To determine the number of significant digits in a number, follow the rules in Section 2.3. Zeros seem to be the most trouble for students. When the measurement is exactly on the calibration line the estimated digit is zero. The words "significant" and "important," for most cases, do not mean the same thing for measured values. A terminal zero is often either significant or important. There are only a few instances where it will be both. If you remove a terminal zero and the value of the number is not changed, then that zero is there only because it is significant. If you remove a terminal zero and the value of the number is changed, then it is there to hold the place value of the number. It is definitely important but it may or may not be significant. You may see significant terminal zeros indicated by a decimal point. For example, 500. would indicate 3 significant digits. 500 would indicate one significant digit. Another method of indicating significant zeros is to put a line over them. Probably the most acceptable solution comes in the use of scientific notation to express measured values. **Every** digit in the number's coefficient in scientific notation is significant. Another common way to express significant digits is to select an appropriate unit in the metric system. For example, if you wanted to show that 500 cm has 2 significant digits, you can report your value as 5.0 m. Your instructor should define the convention to be used in your course.

Calculations involving significant digits can most easily be handled if you remember that *a chain is only as strong as its weakest link.* In addition and subtraction, only the last full column to the right is used for the determination of significant digits in the result. In multiplication and division, the factor with the smallest number of significant digits determines the number of significant digits in the answer.

☐ Read carefully the rules for rounding off a number. Although these are the most common, you may find that your instructor uses slightly different rules. Be sure you understand which rules will be used.

☐ Make sure to read the Examples and work the Practice Problems in the text as you read this section. Check your answers carefully.

☐ Work the Problems for this section at the end of the chapter.

Section 2.4: Exponential Numbers

Some numbers you will be using in your study of chemistry are extremely small and some are extremely large. It would be cumbersome to write out all the zeros to show the size of some numbers. Scientists have devised a system by which these numbers can be expressed more simply. Exponential notation is an important tool. You will want to understand how powers of ten relate to the number of zeros preceding or following a decimal point in an ordinary number. In addition, you must be able to enter and read the exponential numbers from your calculator correctly.

☐ Read Section 2.4 carefully. Note how you are instructed to enter a number in scientific notation into your calculator. The EE or EXP key on the calculator should be read as "times ten to the." If you say this as you enter a number, you will not press any extra keys. Appendix 1 in the back of your textbook tells about calculators and scientific notation. If you have never used scientific notation, you should read this section before you continue with Chapter 2.

☐ If you have not already done so, read Appendix 1 at this time and then return to Section 2.4.

Changing the form of exponential numbers can be simplified by using the following method:
1. Write the number.
2. Count the number of decimal places moved to make the coefficient contain only one non-zero digit to the left of the decimal point.
3. Recognize whether the decimal point is moved left or right.
4. If you moved to the right, subtract the count from the exponent. If you moved to the left, add the count to the exponent.

Write 123.9×10^{13} in scientific notation.

1. 123.9×10^{13}
2. Move decimal 2 positions   (1.239 has only one non-zero digit to the left of the decimal point).
3. This move is made to the left.
4. $1.239 \times 10^{13+2} = 1.239 \times 10^{15}$

Write 0.00657×10^{-3} in scientific notation.

1. 0.00657×10^{-3}
2. Move decimal 3 positions (6.57 has only one non-zero digit to the left of the decimal point).
3. This move is made to the right.
4. $6.57 \times 10^{-3-3} = 6.57 \times 10^{-6}$

To help you recall which way to go, remember that both **R**ight and subt**R**act have an "R." So if you move the decimal to the right you subtract. If you move it to the left, you add.

This method also works when you are writing numbers from ordinary notation to scientific notation. Write the number and add $x\ 10^0$ after the number. By writing $x\ 10^0$ after the number you are not changing the value of the number, since $10^0 = 1$.

Write 65,349 in scientific notation.

1. $65,349 \times 10^0$
2. Must move decimal 4 places.
3. This move is made to the left.
4. $6.5349 \times 10^{0+4} = 6.5349 \times 10^4$

Write 0.000000034500 in scientific notation.

1. $0.000000034500 \times 10^0$
2. Must move decimal 8 places.
3. This move is made to the right.
4. $3.4500 \times 10^{0-8} = 3.4500 \times 10^{-8}$

☐ Finish reading Section 2.4. Since most calculators handle scientific notation you will want to make sure that you can use that function on your calculator. A word to the wise is in order at this point. A calculator only performs those functions the operator tells it to do. If you inadvertently enter an operation or a number improperly, an error will result in your final answer. It is most important to be able to estimate answers to check your calculator. Do not fall into the trap of always relying on your calculator.

☐ Make sure to read the Examples and work the Practice Problems in the text as you read this section. Check your answers carefully.

☐ Work the Problems for this section at the end of the chapter.

Section 2.5: Density

☐ Read Section 2.5. Density is mass divided by volume. Most introductory chemistry courses include a laboratory exercise on density. One important application of density is the determination of whether or not something will float. If two liquids are immiscible, the one with the lower density will be the upper layer. If a solid is placed in a liquid, the solid will float if it has lower density than the liquid. Water is one of the few substances whose solid form is less dense than its liquid form. Wouldn't it be strange to see ice sink to the bottom of a glass of water? For most other substances, however, the solid form is more dense than the liquid form and the solid will not float on top of its liquid.

☐ Make sure to read the Examples and work the Practice Problems in the text as you read this section. Check your answers carefully.

☐ Work the Problems for this section at the end of the chapter.

Section 2.6: Time, Temperature, and Energy

You have been measuring time all of your life. The metric prefixes are used with fractions of seconds as they are with the rest of the metric system. Milliseconds, microseconds, and nanoseconds are commonly used in scientific measurements.

Temperature is also a common measurement. You take your temperature to monitor your health and look at a thermometer to determine what outer clothing to wear as the days get colder or warmer. In chemistry, as in most other sciences, the Celsius (or Centigrade) scale is used. There are exactly 100 Celsius degrees between the normal melting and boiling points of water. There are 180 Fahrenheit degrees between the normal melting and boiling points of water. This means that a Celsius degree is 1.8 times larger than a Fahrenheit degree. You should practice using the conversion equations if your instructor requires you to be able to do this. Most of the time you will need to know quickly how hot something is. For this reason, it is advisable to memorize five approximations between the two scales. These will also give you a way to check your answers when performing the conversions.

Approximate Temperature Conversions

Fahrenheit	Centigrade	
212	100	normal boiling point of water
150	66	hot cup of coffee
99	37	approximate body temperature
72	22	room temperature
32	0	freezing point of water

At what temperature do the Fahrenheit and Centigrade scales read the same?

In this problem, C = F

$$(5/9)(F - 32) = F$$
$$5F - 160 = 9F$$
$$-160 = 4F$$
$$-40 = F$$

Therefore, -40°F is the same as -40°C.

Energy is a fundamental topic in chemistry. The most important thing to remember about energy is that the lowest energy of a system is usually favored. Look back to Chapter 1 and remember some of the forms of energy in Table 1.3. Some of the units of energy are familiar, such as the calorie. One calorie is the amount of energy necessary to raise the temperature of one gram of water one degree Celsius. Note that the nutritional Calorie, used as an energy unit in metabolism, is actually a kilocalorie. The joule is the SI unit for energy. An explanation of kinetic energy ($\frac{1}{2}mv^2$) is necessary to understand the units of joules. Mass times the square of the velocity will have the units of kg · (meters/second)^2, or

$$kg \cdot m^2 / s^2$$

☐ Make sure to read the Examples and work the Practice Problems in the text as you read this section. Check your answers carefully.

☐ Work the Problems for this section at the end of the chapter.

Chapter 2: Finishing Up

☐ Carefully read the summary section at the end of the chapter. Do you understand each paragraph? Do you know the terms used? If not, review the section indicated at the end of the paragraph.

☐ The section titled "Items for Special Attention" gives you some pointers to help you learn the material correctly and also can keep you from making some of the more common mistakes. Read these carefully.

☐ Work the "Self-Tutorial Problems." Check each answer as you complete the problem. If you have trouble with one of these, be sure to ask your instructor for help.

☐ Look over the list of "Key Terms" at the beginning of the chapter. If you do not recognize a term or are unsure of how it was used in the chapter, go back to that section and reread it.

☐ Use your lecture notes and the text to find the important topics covered in class. Make flash cards so that you can study these carefully before the exam.

☐ Work each of the problems at the end of the chapter in your text. Be sure to check your answers. If the answer is not given for that problem, work another that is similar and check the answer. If you are incorrect, check your work and review the chapter to see if you can answer the question correctly. If you cannot, get help from your instructor.

☐ Work the sample exam questions on the next few pages.

☐ Check your answers when you have completed all of them.

☐ Make up your own exam and exchange it with a fellow student or use it in your study group.

Chapter 2: Sample Exam Questions

1. You are preparing to take a troop of scouts camping. In planning for the evening meal you need to know how many pounds of hamburger to order. For your family of six you would prepare the meal using 2.5 pounds of hamburger. If there are 32 scouts (including the adults), how much hamburger should you order?

2. How many meters are there in 2.54 cm?

3. How many grams are there in 454 kg?

4. Convert each of the following to scientific notation:

    (a) 35,781 _____

    (b) 0.0000000235 _____

    (c) 23.65×10^4 _____

    (d) 0.0009872×10^{-5} _____

    (e) 5,781,000 _____

    (f) 0.00006892 _____

    (g) 146.87×10^6 _____

    (h) $0.00000008312 \times 10^{-10}$ _____

5. What is the density of a bar of soap if it is 3.12 cm x 4.45 cm x 1.09 cm and has a mass of 7.67 grams?

6. Will the bar of soap in question 5 float on water before it dissolves?

7. Perform the following conversions within the metric system:

    (a) 350 km = _____ m

    (b) 9.814 mL = _____ cm^3

    (c) 0.00675 m = _____ mm

    (d) 2.13 mL = _____ L

    (e) 27 mg = _____ µg

8. The earth contains approximately 7.5% aluminum by mass. If the earth weighs 3×10^{24} grams, how many kilograms of aluminum are present?

9. The radius of a copper atom is 1.28 angstrom (Å). A copper atom weighs 1.06×10^{-22} g. (1 Å = 1×10^{-10} meter)

    (a) What is the volume of a copper atom in cm^3?
        [$V = (4/3)\pi r^3$]

    (b) What is the density of a copper atom in g/cm^3?

10. To determine the density of an unknown metal a student measured the volume of water in a graduated cylinder. He then weighed a piece of metal and carefully placed the metal in the cylinder with the water. He then recorded the new level of the water.

        volume of water   24.73 mL

        mass of metal   5.72 g

        new volume   26.85 mL

What is the density of the metal? Which of the metals in Table 2.5 of your text do you think the student was assigned?

11. Oil spreads on water to form a film that is about 0.1 µm thick. If a tanker leaks 500 barrels of oil (1 barrel = 31.5 gallons), how many square miles of ocean will be covered with the oil slick? [Hint: area = length x width, volume = length x width x height, therefore, area = volume/ thickness (height).]

12. The diameter of a red blood cell is about 6 µm. If cells were placed in a line, how many red blood cells would it take to make one centimeter?

Chapter 2: Experiment at Home

Introduction to Thermodynamics

Almost all chemical reactions either absorb or release energy in the form of heat. Thermodynamics is the study of the relationship between heat and other forms of energy.

The best way to understand heat energy is to perform a few simple experiments. Prepare three containers into which you can place your hand. In the first container place warm water, into the second place room temperature water and into the third place ice water. Start the experiment by placing one hand in the warm water and the other into the ice water. After a few minutes, place both hands in the room temperature water The hand from the hot water feels cold and the hand from the cold water feels warm. This is caused by the gain or loss of heat from your hand. Take a moment and try this experiment.

Heat energy is also transferred when liquid is evaporated from the skin. When a wet cloth is placed on your forehead you can feel the coolness. The evaporation of the water from the skin requires a certain amount of energy so that the water can go from the liquid to the gas phase. The heat for this vaporization is absorbed from your skin. This energy transfer results in the cooling of your feverish brow. Alcohol is also used to cool. Alcohol evaporates faster, and despite the fact that the alcohol will feel cooler in the beginning, the water will result in better total cooling because it will absorb more energy in the vaporization process.

Chapter 3: Atoms and Atomic Masses

Before you get started

You will need to understand percentages. If you find that you are having difficulties with these calculations, find a mathematics workbook that gives directions for using decimals and percentages. You will need to practice so that you can understand the concepts of this chapter.

Section 3.1: Laws of Chemical Combination

Once Lavoisier convinced other chemists that mass was conserved in a chemical reaction, it was possible for other quantitative relationships to be made. If mass is conserved in a chemical reaction, then atoms must also be conserved.

The **law of definite proportions** states that a compound, regardless of its origin or method of preparation, always contains the same elements in the same proportions by weight; its composition is constant.

The **law of multiple proportions** states that when two elements combine to form more than one compound, the mass of one element which combines with a fixed mass of the other element is in a ratio of small whole numbers. For example, NO, NO_2, and N_2O_3 are all compounds of nitrogen and oxygen and for a given mass of nitrogen have masses of oxygen in a ratio of 2:4:3; that is, for every two atoms of nitrogen there are 2, 4 or 3 atoms of oxygen.

The percent composition can be calculated for each element in a compound.

$$\% \text{ of element} = \frac{\text{mass of element}}{\text{mass of compound}} \times 100$$

The word "percent" often gives students trouble. It is actually an abbreviation for "per one hundred." Anytime you see a number written as a percentage, just think to yourself that this is the fractional part of one hundred.

$$25\% = 25/100 = 0.25$$

$$50\% = 50/100 = 0.50$$

$$100\% = 100/100 = 1.00$$

7.15% of your paycheck is deducted for social security. If your gross pay is 1000 dollars per week, how much do you get in your net pay?

$$\$1000 \times \frac{\$7.15 \text{ deducted}}{\$100 \text{ earned}} = \$71.50$$

or $7.15\% = 7.15/100 = .0715$

$$\$1000 \times .0715 = \$71.50$$

so $\$1000 - \$71.50 = \$928.50$

Another way of looking at the same problem is to recognize that the answer you need is what is left. Whenever percentages are given, the total of all items will always be 100%. If you are deducting 7.15% from the paycheck, then you are taking home 100% - 7.15%, or 92.85%.

$$\$1000 \times \frac{\$92.85 \text{ take home}}{\$100 \text{ earned}} = \$928.50$$

$$\$1000 \times 0.9285 = \$928.50$$

Money is counted to pennies. Significant digit rules are not exactly obeyed because of this. When you fill out your tax forms you are given the option of rounding to whole dollar amounts, but for most circumstances money is counted to the nearest penny.

☐ Read carefully Section 3.1. The law of multiple proportions simply states for a given mass of one element the ratio of masses of the other element(s) is a small whole number ratio. Example 3.3 illustrates the fact that when the amount of carbon is constant, the amount of oxygen in the carbon dioxide is two times the amount of oxygen in the carbon monoxide.

☐ Make sure to read the Examples and work the Practice Problems in the text as you read this section. Check your answers carefully.

☐ Work the Problems for this section at the end of the chapter.

Section 3.2: Dalton's Atomic Theory

☐ Read Section 3.2. Why was it so much easier to understand the laws of chemical combination using the atomic theory?

☐ Reread the five postulates at the beginning of this section. Write each one in your own words below:

1. _____

2. _____

3. _____

4. _____

5. _____

Which of the postulates is entirely incorrect?

☐ Make sure to read the Examples and work the Practice Problems in the text as you read this section. Check your answers carefully.

☐ Work the Problems for this section at the end of the chapter.

Section 3.3: Subatomic Particles

There are some basic properties that need to be mentioned before you learn about the particles.

1. Some matter has charge. It can be either positive or negative.

2. When samples of matter of the same charge are close, they repel each other. If the charges are opposite, the samples attract each other.

☐ Memorize Table 3.1. Your instructor may have you round off the mass to the nearest whole number. If this is done, the proton has a mass of one, the neutron has a mass of one and the electron has a mass of zero. This zero mass simply means that when adding the masses of the particles, the mass of the electron is insignificant with respect to the mass of the other two particles.

The following words which appear in this section are very important. Use your own words to define each of these:

nucleus _____

neutral _____

atomic number (Z) _____

isotope _____

mass number (A) _____

Remember that often when writing mathematical expressions, symbols and numbers which are not necessary to understanding are deleted. When you write the symbols for these isotopes it is understood that the atomic number is determined by the symbol for the element. It is unnecessary to repeat this number. Some of the important isotopes you will learn about in Chapter 19 are cobalt-60, iodine-131, and carbon-14. Cobalt-60 has 27 protons, 33 neutrons, and 27 electrons.

Iodine-131 has _____ protons, _____ neutrons, and _____ electrons.

Carbon-14 has _____ protons, _____ neutrons, and _____ electrons.

☐ Make sure to read the Examples and work the Practice Problems in the text as you read this section. Check your answers carefully.

☐ Work the Problems for this section at the end of the chapter.

Section 3.4: Atomic Mass

The atomic mass of an element is actually the weighted average of the masses of all of the naturally occurring isotopes of that element. Isotopes are atoms with the same number of protons but with a different number of neutrons. Most elements occur as mixtures of isotopes. For this reason, the atomic mass of an element is usually not a whole number. It is the average mass of all naturally occurring isotopes, taking into account the relative abundance of each one. This mixture remains constant throughout nature.

☐ Make sure to read the Examples and work the Practice Problems in the text as you read this section. Check your answers carefully.

☐ Work the Problems for this section at the end of the chapter.

Section 3.5: Development of the Periodic Table

The recognition that properties of the elements repeated in an orderly fashion is still a remarkable feat. Mendeleyev certainly deserves a great deal of admiration for his accomplishments. The significance of the periodic table will become better understood as you learn more in the following chapters.

Look at the periodic table in the front of your book. The elements are arranged by increasing atomic number. The text gives you one example of two elements that did not fit when placed according to increasing atomic weight. Can you find any others?

A periodic table will have at least three items in each box. Many periodic tables include much more information, but we will just start with the basics. The number which is always a whole number (integer) is the number of protons (the atomic number) of the element. The symbol for the element is in the middle. At the bottom of each box is the atomic mass for the element. Some of the elements at the bottom of the table have only unstable isotopes. The whole number appearing in parentheses is the mass number of the most stable of the isotopes.

☐ Make sure to read the Examples and work the Practice Problems in the text as you read this section. Check your answers carefully.

☐ Work the Problems for this section at the end of the chapter.

Chapter 3: Finishing Up

☐ Carefully read the summary section at the end of the chapter. Do you understand each paragraph? Do you know the terms used? If not, review the section indicated at the end of the paragraph.

☐ The section titled "Items for Special Attention" gives you some pointers to help you learn the material correctly and also can keep you from making some of the more common mistakes. Read these carefully.

☐ Work the "Self-Tutorial Problems." Check each answer as you complete the problem. If you have trouble with one of these, be sure to ask your instructor for help.

☐ Look over the list of "Key Terms" at the beginning of the chapter. If you do not recognize a term or are unsure of how it was used in the chapter, go back to that section and reread it.

☐ Use your lecture notes and the text to find the important topics covered in class. Make flash cards so that you can study these carefully before the exam.

☐ Work each of the problems at the end of the chapter in your text. Be sure to check your answers. If the answer is not given for that problem, work another that is similar and check the answer. If you are incorrect, check your work and review the chapter to see if you can answer the question correctly. If you cannot, get help from your instructor.

☐ Work the sample exam questions on the next few pages.

☐ Check your answers when you have completed all of them.

☐ Make up your own exam and exchange it with a fellow student or use it in your study group.

Chapter 3: Sample Exam Questions

You will need a periodic table.

1. Complete the following table for neutral atoms of specific isotopes:

	Isotope	Atomic Number	Mass Number	No. of Protons	No. of Neutrons	No. of Electrons
(a)		25	55			
(b)			18			8
(c)		13			14	
(d)				12	13	
(e)		10	20			
(f)			11			5
(g)		24			30	
(h)				26	28	

2. What is the atomic mass of sulfur? The naturally occurring isotopes are

Isotope	Percentage of Natural Abundance	Relative Mass (amu)
^{32}S	95.0%	31.97207
^{33}S	0.76%	32.97146
^{34}S	4.22%	33.96786
^{36}S	0.014%	35.96709

3. What is the atomic mass of zinc? The naturally occurring isotopes are

Isotope	Percentage of Natural Abundance	Relative Mass
^{64}Zn	48.89%	63.9291
^{66}Zn	27.81%	65.9260
^{67}Zn	4.11%	66.9271
^{68}Zn	18.57%	67.9249
^{70}Zn	0.62%	69.9253

4. In a certain experiment 32.89 grams of a compound containing potassium, chlorine and oxygen is heated releasing the oxygen. If the mass of the compound after heating is 20.00 grams, what was the percentage of oxygen in the original sample?

5. Fill in the blanks in the following table for neutral atoms:

	Isotope	Atomic Number	Mass Number	No. of Protons	No. of Neutrons	No. of Electrons
(a)			209	83		
(b)			89			39
(c)					4	3
(d)		59	141			
(e)				30	34	

6. Calculate the atomic mass of silicon if 92.21% of naturally occurring silicon atoms have a mass of 27.97693 amu, 4.70% have a mass of 28.97649 amu, and 3.09% have a mass of 29.97376 amu.

7. In the first column you are given a known compound. Based on your knowledge of periodic trends, predict the formula for a compound between the two elements listed in the second column. Write your answer in the third column.

H_2O	hydrogen and sulfur	
NaCl	potassium and chlorine	
CH_4	silicon and hydrogen	
MgO	calcium and oxygen	
$ZnCl_2$	cadmium and chlorine	

8. According to Dalton's Atomic Theory:

    (a) matter is made up of very tiny, indivisible particles called _____.

    (b) atoms combine to form _____.

    (c) when atoms combine they do so in _____-_____ ratios.

    (d) the atoms of each element all have the same _____.

9. Two compounds of carbon and hydrogen have the following compositions. Show that these compounds obey the law of multiple proportions.

	percent carbon	percent hydrogen
Compound 1	85.71	14.29
Compound 2	92.31	7.69

10. Fill in the following table about subatomic particles:

Name	Charge	Mass	Location
	1-		outside nucleus
proton		1	
	0		

Chapter 3: Experiment at Home

Visualizing Emission Spectra

When atoms become electronically excited or selectively ionized, a reverse of the phenomenon shown in Figure 3.7 of the text is sometimes seen. The atoms give off light at very specific wavelengths. This occurs in mercury vapor lamps and sodium vapor lamps and in a variety of outdoor display lights.

Materials:
    1 empty facial tissue box
    1 pair scissors
    1 roll opaque tape
    1 CD disk

Method: Make a hole approximately one centimeter square in the middle of the small end of a facial tissue box. Next, mask off most of this hole using opaque tape so as to create a narrow slit, approximately one millimeter wide, parallel to the bottom of the box.

On the bottom of the box, at the opposite end, make a cut across the full width of the box. Insert the CD through this cut until it touches the top of the box. Use tape to secure the CD flat to the end wall of the box. Additional tape should be placed around the cut to prevent stray light entry. You now have created a simple spectroscope.

To see the diffraction of colors, place the box next to your face and look at the CD through the hole from which you would normally remove tissues. The light to be examined should be at your back and must enter the tissue box through the small slit. The diffraction pattern will be seen as the light is reflected from the CD to the large opening in the tissue box. You may need to adjust the size of the slit to get the best division of colors. Examine a fluorescent lamp. The colors that make up this white light are often yellow, green, blue, and purple. Try different types of lighting. After some investigation you should discover that incandescent lamps show a continuum of colors while most other lights emit light at only a few discrete wavelengths.

Theory: Very hot objects, like the sun or an incandescent lamp, often give off a continuum of colors, whereas if the substance can be selectively excited, then only very thin lines of emission are seen. This latter type of emission is typical of situations where the electrons are tightly confined to the atom or molecule but are selectively excited between different electronic or orbital states. By contrast, the electrons on the sun are essentially free to move between a large variety of energy states such that a continuum of colors is seen. An improved spectrum analyzer can be built using a precision optical diffraction grating. Several different quality gratings can be obtained from a scientific supply house.

WARNING: NEVER LOOK DIRECTLY AT THE SUN.

Chapter 4: Electronic Configuration of the Atom

Before you get started

Formation of compounds and chemical reactions involve the interaction of the electrons of one atom with those of others. Therefore, it is useful to know the arrangement of the outer electrons in an atom. To do this you will be learning some rules about electronic configuration.

Remember the properties of protons, neutrons and electrons that you learned in Chapter 3. In a neutral atom, the number of electrons is always equal to the number of protons. This number is easily found by looking up the element in the periodic table.

Section 4.1: Bohr Theory

Seats at a sporting event are useful as a model for the discrete energy levels occupied by electrons in an atom. If electrons are spectators, then you can assume that the lowest seats are the "best" seats, or the ones that will always be claimed first. A spectator would want to always sit in the lowest seat possible. The concept of quantized energy can easily be explained by illustrating that no one can sit between seats. No two spectators can have exactly the same seat at the same time. Each spectator is assigned a combination of section, row, seat, and date that is different from that of every other spectator.

When all of the electrons are packed into the lowest possible configuration, this is called the ground state. If an electron is energized into a position higher than its ground state, it is called an excited state. As the electron "leaps" back to the ground state, energy equal to the difference in positions is emitted. An example of emission spectroscopy can be found in street lamps. The yellow lamps are sodium vapor lamps. Blue-white lamps are mercury vapor lamps. What you are actually observing when you see the light is the energy being emitted as electrons return to the ground state. The amount of energy is different for each element. The take-home lab at the end of this chapter will tell you some interesting facts about fireworks based on the characteristic emission spectra of several elements.

☐ Make sure to read the Examples and work the Practice Problems in the text as you read this section. Check your answers carefully.

☐ Work the Problems for this section at the end of the chapter.

Section 4.2: Quantum Numbers

The four quantum numbers are used to identify each electron in the atom. Every electron has its own set of numbers that tell the electron's level, sublevel, type of orbital, and which direction the electron is spinning.

☐ Read Section 4.2.

Table 4.1 gives the guidelines for assigning the permitted values to an electron. These should be memorized before you continue.

☐ Make sure to read the Examples and work the Practice Problems in the text as you read this section. Check your answers carefully.

☐ Work the Problems for this section at the end of the chapter.

Section 4.3: Energies of Electrons

☐ Read Section 4.3.

In order to properly place the electrons into orbitals you need to know the relative energies of the orbitals. The *n + l rule* can be used to establish the order of increasing energy of the electron positions.

1. Assign the quantum numbers *n* and *l*.

2. The larger the sum of *n* and *l*, the higher the energy of the orbital.

3. If the sums of two sets of *n* and *l* are the same, then the orbital with the higher *n* value has the higher energy.

4. If two electrons are in orbitals with the same value of *n* and the same value of *l*, then the electrons are said to be in degenerate orbitals – orbitals of the same energy.

☐ Make sure to read the Examples and work the Practice Problems in the text as you read this section. Check your answers carefully.

☐ Work the Problems for this section at the end of the chapter.

Section 4.4 Shells, Subshells, and Orbitals

☐ Read Section 4.4. Read the definitions very carefully in the first paragraph of this section. Define the following in your own words:

shell: _____

subshell: _____

orbital: _____

The easiest way to memorize these is to remember what is the same in each case.

> All of the electrons in the same **shell** have the same n quantum number. The other quantum numbers may be different.
>
> All of the electrons in the same **subshell** have the same n and l quantum numbers.
>
> Both electrons in the same **orbital** have the same n and l and m quantum numbers.

The type of subshell corresponding to each of the l values is important to remember. You will also need to remember how many electrons are permitted in each of the subshells. For example, when $n = 1$ and $l = 0$ the subshell is called $1s$. When $n = 2$ and $l = 0$, the subshell is called $2s$. For $n = 2$ and $l = 1$, the subshell is called $2p$. An s subshell can hold a maximum of two electrons. A p subshell can hold a maximum of six electrons. A d subshell can hold a maximum of ten electrons. And, an f subshell can hold a maximum of fourteen electrons. The chart on the following page may be helpful.

This should be enough to get you started. The $n + l$ rule will enable you to write the increasing order of the subshells. If you also remember the maximum number of electrons that are allowed in each of the subshells, you will be able to write the electronic configuration for any element in its ground state.

☐ Can you identify the subshell for the n and l values in Example 4.8?
    (a) $4p$ (b) $4s$ (c) $3d$ (d) $3p$

☐ Make sure to read the Examples and work the Practice Problems in the text as you read this section. Check your answers carefully.

☐ Work the Problems for this section at the end of the chapter.

n	l	m							name of subshell	maximum number of electrons permitted
1	0				0				1s	2
2	0				0				2s	2
	1			-1	0	+1			2p	6
3	0				0				3s	2
	1			-1	0	+1			3p	6
	2		-2	-1	0	+1	+2		3d	10
4	0				0				4s	2
	1			-1	0	+1			4p	6
	2		-2	-1	0	+1	+2		4d	10
	3	-3	-2	-1	0	+1	+2	+3	4f	14
5	0				0				5s	2
	1			-1	0	+1			5p	6
	2		-2	-1	0	+1	+2		5d	10
	3	-3	-2	-1	0	+1	+2	+3	5f	14
6	0				0				6s	2
	1			-1	0	+1			6p	6
7	0				0				7s	2

Section 4.5: Shapes of Orbitals

Orbitals are defined by mathematical equations. The probability of finding an electron in space is calculated, and the volume of the highest probability is defined as the orbital. The formation of bonds between atoms depends on how their orbitals overlap.

☐ Read Section 4.5.

☐ Make sure to read the Examples and work the Practice Problems in the text as you read this section. Check your answers carefully.

☐ Work the Problems for this section at the end of the chapter.

Section 4.6: Energy Level Diagrams

Energy level diagrams are constructed based on the increasing energy of the subshells. Note that in Figure 4.6 the degenerate levels (those with the same n and l and m quantum numbers) are placed in the same horizontal level. Also note that when the $n + l$ **rule** is applied to increasing energy, the $4s$ is lower in energy than the $3d$.

☐ Look over Figure 4.6. Using the $n + l$ **rule** determine the value of $n + l$ for the orbitals given and place them in order below..

	$n+l$		$n+l$		$n+l$		$n+l$
$1s$	_____	$3d$	_____	$5s$	_____	$6p$	_____
$2s$	_____	$4s$	_____	$5p$	_____	$6d$	_____
$2p$	_____	$4p$	_____	$5d$	_____	$7s$	_____
$3s$	_____	$4d$	_____	$5f$	_____	$7p$	_____
$3p$	_____	$4f$	_____	$6s$	_____		

Order of filling: ____, ____, ____, ____, ____, ____, ____, ____, ____, ____, ____,

____, ____, ____, ____, ____, ____, ____, ____,

☐ Read Section 4.6 at this time. One rule that has increased the understanding of magnetic behavior is Hund's rule, which states that all degenerate orbitals will fill with single electrons before the second electron is placed in any of these.

☐ Make sure to read the Examples and work the Practice Problems in the text as you read this section. Check your answers carefully.

☐ Work the Problems for this section at the end of the chapter.

Section 4.7: Periodic Variation of Electronic Configuration

☐ Fill in the periodic table on the next page to show the subshell of the outermost electron for Li, Na, and K. Do the same for F, Cl and Br. Check your answers with Figure 4.8. You can predict the subshell of the outermost electron based on the position of that element in the periodic table.

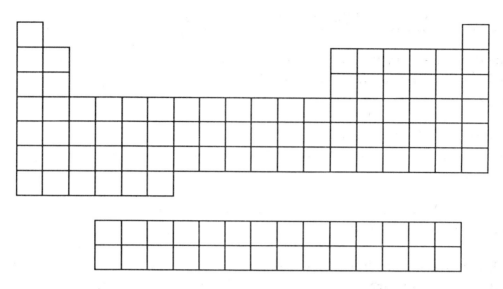

☐ Look back at Figure 4.6. Compare the values you calculated for $n + l$ for each of the subshells to the order of Figure 4.8.

Shortened notation is often used since you will be most concerned with the outermost electrons. Look back to Practice Problem 4.14 and use the electronic configurations of the noble gases to write the shortened notation for the elements in Example 4.15.

(a) P   [Ne] $3s^2 3p^3$

(b) Fe [Ar] $4s^2 3d^6$

(c) La [Xe] $6s^2 5d^1$

☐ Make sure to read the Examples and work the Practice Problems in the text as you read this section. Check your answers carefully.

☐ Work the Problems for this section at the end of the chapter.

Chapter 4: Finishing Up

☐ Carefully read the summary section at the end of the chapter. Do you understand each paragraph? Do you know the terms used? If not, review the section indicated at the end of the paragraph.

☐ The section titled "Items for Special Attention" gives you some pointers to help you learn the material correctly and also can keep you from making some of the more common mistakes. Read these carefully.

☐ Work the "Self-Tutorial Problems." Check each answer as you complete the problem. If you have trouble with one of these, be sure to ask your instructor for help.

☐ Look over the list of "Key Terms" at the beginning of the chapter. If you do not recognize a term or are unsure of how it was used in the chapter, go back to that section and reread it.

☐ Use your lecture notes and the text to find the important topics covered in class. Make flash cards so that you can study these carefully before the exam.

☐ Work each of the problems at the end of the chapter in your text. Be sure to check your answers. If the answer is not given for that problem, work another that is similar and check the answer. If you are incorrect, check your work and review the chapter to see if you can answer the question correctly. If you cannot, get help from your instructor.

☐ Work the sample exam questions on the next few pages.

☐ Check your answers when you have completed all of them.

☐ Make up your own exam and exchange it with a fellow student or use it in your study group.

Chapter 4: Sample Exam Questions

1. What values of m are permitted for an electron with $l = 3$?

2. What values of m are permitted for an electron with $l = 1$?

3. What values of l are permitted for an electron with $n = 3$?

4. What values of l are permitted for an electron with $n = 2$?

5. What type of subshell corresponds to $l = 2$?

6. What type of subshell corresponds to $l = 3$?

7. Write the detailed electronic configuration for magnesium.

8. Write the detailed electronic configuration for sulfur.

9. Write the detailed electronic configuration for nitrogen.

10. Write the detailed electronic configuration for calcium.

11. On the periodic table on the next page show where you would find

(a) elements with the outer shell configuration of $ns^2 np^4$.

(b) The elements with the outer shell configuration of ns^2.

(c) The elements with the outer shell configuration of $ns^2 np^5$.

(d) The elements with the outer shell configuration of $ns^2 np^3$.

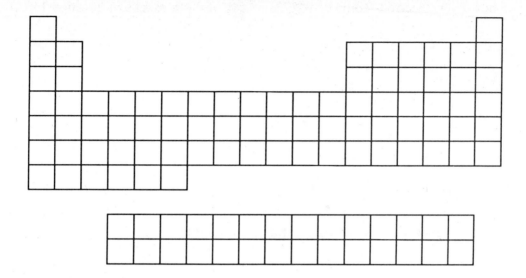

12. Arrange the following electrons, identified by the *n* and *l* quantum numbers, in increasing energy order.

(a) $n = 4, l = 2$ (b) $n = 3, l = 2$ (c) $n = 4, l = 1$, (d) $n = 5, l = 0$

_____ _____ _____ _____
lowest energy   highest energy

13. Write the four quantum numbers for each of the five electrons of boron.

14. Give the name of the subshell that corresponds to each of the following:

(a) $n = 3, l = 2$ _____

(b) $n = 2, l = 1$ _____

(c) $n = 1, l = 0$ _____

(d) $n = 5, l = 3$ _____

(e) $n = 4, l = 3$ _____

15. Tell how many electrons could be placed in each of the subshells of question 14.

    (a) _____

    (b) _____

    (c) _____

    (d) _____

    (e) _____

16. Write the detailed electronic configuration for bromine.

17. Match the theory to the name:

    ___ a. Heisenberg   (1) Electrons in atoms are arranged in shells, each with a definite energy.
    ___ b. Pauli   (2) Electrons within a given subshell remain as unpaired as possible.
    ___ c. Hund   (3) No two electrons in an atom can have the same set of four quantum numbers.
    ___ d. Bohr   (4) It is impossible to know both the momentum and the location of an electron at the same time.

18. Give the designations for the subshells possible for $n = 4$.

19. What is the maximum number of unpaired electrons possible in a p subshell?

20. Write the detailed electronic configuration for calcium.

21. Using noble gas shortened configuration, give the electronic structure of uranium.

22. Into which subshell would the last electron of each of the following be placed:

    (a) Mo _____

    (b) Te _____

    (c) V _____

    (d) Cd _____

    (e) Bi _____

23. Mendeleyev discovered that groups of elements have similar chemical properties. What feature of the group members makes this true?

Chapter 4: Experiment at Home

Fun with Fireworks

Most of us at one time or another have enjoyed a good fireworks display. In this experiment you will learn what makes the beautiful colors and some of the special effects. This is not a how-to-do-it lab and you will not be able to make your own when you are done, but you should have some appreciation for the use of the various elements and their emission spectra in the production of pyrotechnics.

Fireworks have four components: oxidizer, fuel, binder, and special effects. Oxidizers are the substances which cause the fuel to burn. The binder holds everything together. The special effects consist of color, sparkles, smoke, and noise.

Color is the primary effect of fireworks. This property is determined by the element used in the powders. The following list are the common elements that have been used in the industry for centuries.

red	strontium
green	barium
blue	copper
yellow	sodium
white	magnesium

Some of the substances used for sparkle effects are iron filings for gold sparks and aluminum-magnesium alloy for white sparks.

The smoke is usually a sulfur-based material with an organic dye for different colors. Flash powders used at rock concerts are a mixture of potassium perchlorate and magnesium.

Whistle effects are produced by benzoate or salicylate salts. Other loud noises and bangs are explosives designed to detonate at predicted intervals during the flight of the shell.

The next time you see fireworks, remember each color is produced by burning a compound containing a specific element.

Chapter 5: Chemical Bonding

Before you get started

☐ Review the charges of the subatomic particles in Section 3.3.
☐ Remember how to determine the electronic configuration for an atom.
☐ Be able to tell the number of outermost shell electrons in an atom of an element.

You will want to make notes on a periodic table as you read through the chapter. You may make copies of the one below to use as you read your text.

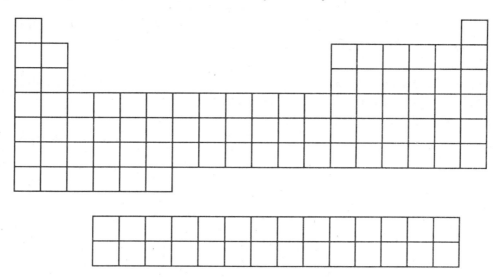

Section 5.1: Chemical Formulas

☐ Read Section 5.1.

☐ Using the periodic table above, locate the most electronegative element.

☐ Locate the most electropositive element.

With this information, you can now predict with some certainty which of any two elements will be more electronegative. The more electronegative element is usually placed to the right in the formula of a binary chemical compound. Note the important exception of hydrogen. Chemists write the H first in HCl and H_2O and last in NH_3. (See rules for acids in Chapter 6)

☐ Locate the elements which form molecules when free (diatomic elements) on the periodic table. Can you find the figure "7" that they form? What other elements occur in polyatomic elemental forms? The seven diatomic elements, elemental phosphorus (P_4), and elemental sulfur (S_8) should be memorized. Remember that these forms occur only when the element is present as a *element*. When these elements combine with other elements to form compounds, they obey other rules, some of which you will learn in this chapter.

☐ Carefully examine the examples in Section 5.1. You will need to decide how many atoms of each element there are in a formula. The use of a subscript after an atomic symbol tells how many atoms of that element there are in a formula unit of that compound. Parentheses mean you must multiply everything inside the parentheses by the following subscript in order to account for all of the atoms.

☐ Cover up the following answers with a sheet of paper and see if you can count the correct number of atoms of each element in the following formulas:

Formula	Answer
$Al(OH)_3$	1 aluminum, 3 oxygen and 3 hydrogen
$Ca(OH)_2$	1 calcium, 2 oxygen and 2 hydrogen
$Fe(NO_3)_2$	1 iron, 2 nitrogen and 6 oxygen

☐ Make sure to read the Examples and work the Practice Problems in the text as you read this section. Check your answers carefully.

☐ Work the Problems for this section at the end of the chapter.

Section 5.2: Ionic Bonding

☐ You will need your periodic table for this section.

☐ Read Section 5.2.

An ion is formed from an atom when there is a net charge present, a result of either removing electrons or adding them. An uncombined atom has the same number of electrons as it has protons in its nucleus. When this is the case, the net charge is zero. If one or more electrons are removed, then there is less negative charge than positive charge, and the ion is positive. If one or more electrons are added, then there is more negative charge than positive charge, and the ion is negative.

The number of electrons added to or subtracted from a neutral atom to form an ion is determined for most main group elements using the **octet rule**. This is a tendency for the atom to gain or lose electrons so that the outermost shell has a total of eight electrons. Metals, on the left side of the periodic table, are most likely to lose electrons and form positive ions. Nonmetals, on the right side of the periodic table are more likely to gain electrons and form negative ions.

☐ Find the main group elements on the periodic table. These are the elements in the first two and the last six columns of the table. For atoms of these elements you can tell how many electrons are in their outer shells by simply looking at the classical group number at the top of each column.

Ions that have a positive charge are called "cations" because they are attracted to the negatively charged cathode. Ions that have a negative charge are called "anions" because they are attracted to the positively charged anode. You should remember these two words and be able to identify the cation and anion part of any ionic substance. The cation is always written first in the formula of an ionic substance.

Note at this time that ionic substances exist in the form of cations and anions in solution. Most molecular substances do not form ions and for this reason they do not conduct electricity. Look at Figure 5.9.

☐ The transition metals with partially full *d* and *f* orbitals obey similar rules; but, they can lose electrons from *s*, *p*, *d* or *f* subshells. Read carefully the paragraphs Detailed Electronic Configurations of Ions in Section 5.2. Be sure to work the practice problems in this section.

Formulas for ionic compounds are deduced by simply making sure that the overall charge on the compound is neutral. The number of positive charges from the cations must equal the number of negative charges from the anions. The numbers of cations and anions in a compound are designated by using a subscript following the symbol (or formula) for that ion.

☐ Many important ions are polyatomic. Examples are the nitrate ion, NO_3^-, the sulfate ion, SO_4^{2-}, and the phosphate ion, PO_4^{3-}. These and several others can be found in Table 5.2. Your instructor should assign the ones that you will need to commit to memory. Be sure to memorize these as soon as possible.

☐ Make sure to read the Examples and work the Practice Problems in the text as you read this section. Check your answers carefully.

☐ Work the Problems for this section at the end of the chapter.

Section 5.3: Electron Dot Diagrams

Electron dot diagrams enable you to visualize the valence electrons on an atom. To construct a diagram you may first imagine a box around the element's symbol. A maximum of two electrons may be placed on each side of the box when the atom is uncombined.

The electron dot diagram for nitrogen contains five dots representing the electrons in the valence shell. Each of the first four would be placed on one of the sides of the square. The fifth electron would need to be paired with one of the original four. It does not matter which one.

$$\cdot \overset{\cdot\cdot}{\underset{\cdot}{N}} \cdot$$

Dot diagrams for ions must show that electrons have been added or removed. An electron dot diagram for an ion may be shown in brackets with the charge of the ion shown outside the brackets.

☐ Make sure to read the Examples and work the Practice Problems in the text as you read this section. Check your answers carefully.

☐ Work the Problems for this section at the end of the chapter.

Section 5.4: Covalent Bonding

☐ Carefully read the first paragraph of this section.

A single covalent bond is formed when an electron pair is shared between two atoms. Be sure you understand the difference between covalent and ionic bonding. You will also see that when two pairs of electrons are shared between the same pair of atoms a double bond is formed. Three pairs of electrons shared between the same pair of atoms form a triple bond. A pair of electrons that is not involved in bonding is called an *unshared pair*, or *lone pair*. One important difference between covalent and ionic bonding is the placement of the electrons in the structure of the compound. In the formation of a simple binary ionic compound, you learned that electrons are transferred from one atom to another to form ions. The electrons are removed from one atom to form a cation and are added to the other atom to form the anion. In covalent molecules, electrons are shared between all of the atoms. There are no ions formed.

☐ Finish reading the section.

It is quite simple to check the electron dot diagrams for accuracy. Just make sure of two things:

1) The number of electrons in the diagram is the same as the number available from the atoms involved. Don't forget to add electrons for negatively charged ions and to subtract electrons for positively charged ions.

2) Every atom has the correct number of electrons around it. For most nonmetals, this will be eight, but there is a major exceptions to this rule, hydrogen has only two electrons.

☐ Look over the examples carefully. Note how the circles have been drawn around the electrons to show that when one or more pairs of electrons are shared, all electrons between the two atoms are counted as being on each atom.

☐ Memorize Table 5.2.

☐ Make sure to read the Examples and work the Practice Problems in the text as you read this section. Check your answers carefully.

☐ Work the Problems for this section at the end of the chapter.

Chapter 5: Finishing Up

☐ Carefully read the summary section at the end of the chapter. Do you understand each paragraph? Do you know the terms used? If not, review the section indicated at the end of the paragraph.

☐ The section titled "Items for Special Attention" gives you some pointers to help you learn the material correctly and also can keep you from making some of the more common mistakes. Read these carefully.

☐ Work the "Self-Tutorial Problems." Check each answer as you complete the problem. If you have trouble with one of these, be sure to ask your instructor for help.

☐ Look over the list of "Key Terms" at the beginning of the chapter. If you do not recognize a term or are unsure of how it was used in the chapter, go back to that section and reread it.

☐ Use your lecture notes and the text to find the important topics covered in class. Make flash cards so that you can study these carefully before the exam.

☐ Work each of the problems at the end of the chapter in your text. Be sure to check your answers. If the answer is not given for that problem, work another that is similar and check the answer. If you are incorrect, check your work and review the chapter to see if you can answer the question correctly. If you cannot, get help from your instructor.

☐ Work the sample exam questions on the next few pages.

☐ Check your answers when you have completed all of them.

☐ Make up your own exam and exchange it with a fellow student or use it in your study group.

Chapter 5: Sample Exam Questions

1. Write the formula for each of the ions in the following ionic compounds.

    $Al_2(SO_4)_3$ _____ _____

    $(NH_4)_3PO_4$ _____ _____

    $Ca_3(PO_4)_2$ _____ _____

    $(NH_4)_2SO_4$ _____ _____

2. Draw the Lewis structure for SO_2.

3. Draw the Lewis structure for SO_3.

4. Draw the Lewis structure for ClO_4^-.

5. Draw the Lewis structure for CO_3^{2-}.

6. Circle the element in the following set with the *highest* electronegativity.

                  Na   Cl   Si   Ca   I

7. Identify each element present and how many atoms of each element are present in one formula unit of $(NH_4)_2SO_4$?

8. What is meant by the *octet rule*? Name one important exception to this rule.

9. Write the electronic configuration for the calcium ion (Ca^{2+}).

10. Determine the formula of the compound of the two elements given:

    (a) magnesium and chlorine

    (b) aluminum and sulfur

    (c) cesium and oxygen

    (d) calcium and oxygen

    (e) lithium and nitrogen

11. Of the following compounds which one(s) would you expect to be covalent? Circle your answer(s)

$$NaCl \quad CsF \quad MgI_2 \quad CO_2 \quad LiBr$$

12. Draw the Lewis structure for $NaNO_3$.

13. Draw the Lewis structure for $CaCO_3$.

14. Identify the ions present in each of the following compounds:

   (a) Na_2SO_4

   (b) $CaCO_3$

   (c) $KC_2H_3O_2$

   (d) KCN

   (e) $LiOH$

15. Complete the following table by placing the proper formula in each box:

	OH^-	SO_4^{2-}	PO_4^{3-}
Al^{3+}			
Ca^{2+}			
NH_4^+			

True and False Questions:

_____ 16. Ionic and covalent bonds can be formed by the sharing of one, two or three pairs of electrons.

_____ 17. Pure compounds that are gases and liquids at room temperature are molecular.

_____ 18. Molecules are formed by joining atoms with ionic bonds.

_____ 19. The cathode is the positive electrode since it attracts cations. The anode is the negative electrode since it attracts anions.

_____ 20. In a chemical formula, the more electropositive element is written first.

Chapter 5: Experiment at Home

Growing Crystals

Some crystals are among the most beautiful structures in nature. Interestingly, the outer shape of a crystal reflects the alignment of the atoms within it. The angles of the faces are determined by the packing. A sodium chloride crystal is cubic. If you look at a crystal of table salt, you will see that the faces form a box. The structure of sodium chloride on the sub-microscopic level is the same. No matter how small the crystal gets, the ions will still form a rectangular solid.

Materials: One drinking glass
white string
pencil
paper clip
water
sugar
spoon
stove or hot plate
pan

Method: Boil enough water to fill the glass. While the water is heating, fill the glass to the top with sugar. Slowly add water to the glass and stir carefully with the spoon until all of the sugar is dissolved. Attach the paper clip to one end of the string and tie the other end around the pencil. Place the string in the glass and roll the string around the pencil until the paper clip is just touching the bottom of the glass. Rest the pencil across the top of the glass. Place the glass in a warm, quiet place for several weeks. As the water evaporates, the crystals will grow. You may want to break the crust that will form on the surface in order to allow the solution to evaporate more easily. If you have been careful not to contaminate your set-up, you can now enjoy old-fashioned rock candy with your friends.

Additional Projects: (1) Alum, which can be purchased at the grocery store, will make large, very clear crystals. Make a saturated solution by heating some water in a pan and, while stirring, add the alum until no more will dissolve. You can tell this because there will be undissolved material at the bottom of the pan. Carefully pour the solution into a drinking glass. As above, tie a string on a pencil and allow it to set in a quiet place until crystals begin to grow. (2) You can try to make sodium chloride crystals in the same way as you made the alum.

Chapter 6: Nomenclature

Before you get started

☐ You should be able to give the symbol for the name and the name for the symbol of the common elements in Figure 1.5.

☐ You should know the term electronegativity and understand how this property determines the ionic or covalent character of the bond between two atoms.

☐ You will need to know the usual charges on the ions of the main group elements.

☐ You will need to know the formulas and charges of the polyatomic ions listed in Table 5.2.

☐ You will need to remember how to write formulas of compounds (Chapter 5).

☐ To begin your study of this chapter, make a copy of the flow chart of Figure 6.3 and the outline in Table 6.6. Use these as you read the chapter. Try to find where each section is described on the flow chart and on the outline.

Section 6.1: Binary Nonmetal-Nonmetal Compounds

☐ Read Section 6.1.

☐ Locate this nomenclature rule on the flow chart of Table 6.6.

The compounds of this type are covalent. They are not named as acids. Several common names are mentioned: NH_3 and H_2O should be familiar to you by now. With the help of your periodic table, find the more electronegative element for each case in Example 6.1. Read the solutions carefully.

☐ You should memorize the prefixes shown in Table 6.1. Note that the prefix *mono* is not used for the first element. Also, when the name contains two adjacent vowels, the first is dropped. Carbon monoxide is **not** named ***mono*carbon mon*oo*xide**.

☐ Make sure to read the Examples and work the Practice Problems in the text as you read this section. Check your answers carefully.

☐ Work the Problems for this section at the end of the chapter.

Section 6.2: Naming Ionic Compounds

☐ Read Section 6.2.

Your text mentions that alums are exceptions. This is because these are compounds containing more than one cation. Potassium aluminum sulfate is the alum used as an astringent and can be purchased at a drug store.

☐ Locate the nomenclature rule for ionic compounds on the outline of Table 6.6.

Once you decide that the compound is ionic, you must then decide whether the cation is constant or variable in its charge. The cations formed from elements in Groups IA and IIA are constant. Aluminum (Al^{3+}), cadmium (Cd^{2+}), zinc (Zn^{2+}), silver (Ag^+) and ammonium (NH_4^+) are also constant. Some polyatomic cations are listed in Table 6.2. These should be memorized.

Most transition metals are variable in charge. Look back at Figure 5.10 to see the charges on common ions. When two or more charges are possible, some guidelines must be made to distinguish one from the other. There are two ways in which this is done. First, you can tell the charge on the metal ion from the name it is given. Iron(II) ion is Fe^{2+}, and iron(III) ion is Fe^{3+}. Secondly, you can tell what the charge on the cation in the compound is by knowing the charge on the anion. Since chloride is Cl^-, then the number of chloride ions in the compound formula tells you the charge on the iron ion. $FeCl_2$ is iron(II) chloride and $FeCl_3$ is iron(III) chloride.

The anion part of the formula is a bit more challenging. If there is just one atom forming the anion its charge is determined by subtracting eight from the group number; and the ending used is *-ide*. Polyatomic anions are very common and most of the ones listed in Tables 6.4 and 6.5 should be memorized.

☐ Make sure to read the Examples and work the Practice Problems in the text as you read this section. Check your answers carefully.

☐ Work the Problems for this section at the end of the chapter.

Section 6.3: Naming Acids and Acid Salts

☐ Read the first paragraph of this section. Note carefully the meaning of ionizable hydrogen atom. It is important to recognize that when the hydrogen appears first in a formula it indicates that the compound is an acid.

☐ Read the section "Naming Acids."

☐ Locate this nomenclature rule on the outline of Table 6.6.

Note the change in name for sulfur and phosphorus compounds.

ion	acid
sulfite, (SO_3^{2-})	sulfurous acid, (H_2SO_3)
sulfate, (SO_4^{2-})	sulfuric acid, (H_2SO_4)
phosphite, (PO_3^{3-})	phosphorous acid, (H_3PO_3)
phosphate, (PO_4^{3-})	phosphoric acid, (H_3PO_4)

Acid salts use the word "hydrogen" in their names to indicate that a hydrogen atom is part of the anion. One common acid salt, baking soda, is sodium hydrogen carbonate, ($NaHCO_3$). The prefix *bi-* is commonly used where one of two ionizable hydrogen atoms remains. Bicarbonate of soda or sodium bicarb are other common names for $NaHCO_3$.

☐ Make sure to read the Examples and work the Practice Problems in the text as you read this section. Check your answers carefully.

☐ Work the Problems for this section at the end of the chapter.

Section 6.4: Hydrates

You can simply name the compound and then using the prefixes from Table 6.1, tell how many water molecules are attached.

Examples of hydrates:	
$AlCl_3 \cdot 6H_2O$	aluminum chloride hexahydrate, used as an antiperspirant
$CaSO_4 \cdot H_2O$	calcium sulfate monohydrate, known commonly as plaster of paris
$MgSO_4 \cdot 7H_2O$	magnesium sulfate heptahydrate Epsom salts

☐ Make sure to read the Examples and work the Practice Problems in the text as you read this section. Check your answers carefully.

☐ Work the Problems for this section at the end of the chapter.

Chapter 6: Finishing Up

☐ Carefully read the summary section at the end of the chapter. Do you understand each paragraph? Do you know the terms used? If not, review the section indicated at the end of the paragraph.

☐ The section titled "Items for Special Attention" gives you some pointers to help you learn the material correctly and also can keep you from making some of the more common mistakes. Read these carefully.

☐ Work the "Self-Tutorial Problems." Check each answer as you complete the problem. If you have trouble with one of these, be sure to ask your instructor for help.

☐ Look over the list of "Key Terms" at the beginning of the chapter. If you do not recognize a term or are unsure of how it was used in the chapter, go back to that section and reread it.

☐ Use your lecture notes and the text to find the important topics covered in class. Make flash cards so that you can study these carefully before the exam.

☐ Work each of the problems at the end of the chapter in your text. Be sure to check your answers. If the answer is not given for that problem, work another that is similar and check the answer. If you are incorrect, check your work and review the chapter to see if you can answer the question correctly. If you cannot, get help from your instructor.

☐ Work the sample exam questions on the next few pages.

☐ Check your answers when you have completed all of them.

☐ Make up your own exam and exchange it with a fellow student or use it in your study group.

Chapter 6: Sample Exam Questions

1. Name the following compounds:

   (a) $Cu(NO_3)_2$ _____

   (b) HNO_2 _____

   (c) PCl_5 _____

   (d) NO_2 _____

   (e) MgO _____

2. Write formulas for the following compounds:

   (a) sodium phosphate _____

   (b) iron(II) fluoride _____

   (c) chromium(III) sulfate _____

   (d) sulfurous acid _____

   (e) dinitrogen tetroxide _____

3. Name the following compounds:

   (a) $CuCl$ _____

   (b) HNO_3 _____

   (c) PCl_3 _____

   (d) NO _____

   (e) MnO _____

4. Write formulas for the following compounds:

    (a) calcium phosphate _____

    (b) iron(III) fluoride _____

    (c) magnesium sulfate _____

    (d) sulfuric acid _____

    (e) nitrogen dioxide _____

5. Name the following compounds:

    (a) PCl_5 _____

    (b) N_2O_5 _____

    (c) IF_7 _____

    (d) BrF_5 _____

    (e) ClO_2 _____

6. Give the names of the following compounds:

    (a) K_2SO_4 _____

    (b) $(NH_4)_2CO_3$ _____

    (c) $NaOH$ _____

    (d) $Mg_3(PO_4)_2$ _____

    (e) $AgNO_3$ _____

7. Give the formulas for each of the following compounds:

   (a) magnesium carbonate _____

   (b) sodium chromate _____

   (c) potassium permanganate _____

   (d) ammonium acetate _____

   (e) sodium dichromate _____

8. Give the name for each of the following compounds:

   (a) $Cu(OH)_2$ _____

   (b) Ag_2SO_4 _____

   (c) $PbCl_4$ _____

   (d) $HgCl_2$ _____

   (e) Fe_2O_3 _____

9. Give the formula for each of the following compounds:

   (a) nickel(II) acetate _____

   (b) gold(III) chloride _____

   (c) manganese(IV) oxide _____

   (d) tin(II) nitrate _____

   (e) copper(I) sulfide _____

10. Give the name for each of the following acids:

   (a) HClO$_3$ _____

   (b) H$_2$SO$_4$ _____

   (c) HBrO$_2$ _____

   (d) HIO _____

   (e) H$_3$PO$_3$ _____

11. Give the formulas for each of the following acids:

   (a) sulfurous acid _____

   (b) nitric acid _____

   (c) periodic acid _____

   (d) bromous acid _____

   (e) hypophosphorous acid _____

12. Write the names for the following compounds:

   (a) CdSO$_4$·7H$_2$O _____

   (b) CoF$_2$ _____

   (c) Sn(CrO$_4$)$_2$ _____

   (d) SeO$_3$ _____

   (e) HClO _____

13. Write the formulas for the following compounds:

    (a) zinc chromate   _____

    (b) magnesium permanganate   _____

    (c) lead(IV) acetate   _____

    (d) hydrochloric acid   _____

    (e) sodium chloride   _____

Chapter 6: Experiment at Home

Household Nomenclature

In this experiment you will be doing some exploring through the house. As you read labels, you should be able to recognize names that you have learned in this chapter.

Sodium

Sodium compounds have been associated with high blood pressure. It is now advisable to recognize those foods that contain sodium ions. Look through the labels on the foods you eat and see just how many *surprise* sources of sodium ions you find. Be sure to check vitamin tablets. Write these down for future reference.

Chapter 7: Formula Calculations

Before you get started

☐ Review how to enter numbers in exponential notation into your calculator and how to perform operations correctly using exponential notation.

☐ Be able to work simple percentage problems. If you had trouble in Chapter 3, you may want to get a copy of a mathematics workbook that will help you master this topic.

☐ Review what you learned about writing formulas in Chapter 5

☐ Review what you learned about naming these compounds in Chapter 6.

☐ Have a copy of the periodic table handy as you read this chapter.

Section 7.1: Formula Masses

There are three types of masses you will be using in this section. Atomic mass is the mass of each atom, taken from the periodic table. Formula mass is the sum of the atomic mass of each atom in one formula unit of any substance. Molecular mass is the sum of the atomic mass of each atom in a molecule of a covalent compound.

☐ Check with your instructor to find out how many significant figures you should use from the atomic masses on the periodic table when doing calculations in this course.

☐ Make sure to read the Examples and work the Practice Problems in the text as you read this section. Check your answers carefully.

☐ Work the Problems for this section at the end of the chapter.

Section 7.2: Percent Composition

☐ Read Section 7.2 carefully. Be sure you understand percentages as used in this section. The easiest way to explain this concept is to say that the percent composition of a particular element in a compound is the mass of all atoms of that element that would be found in 100 grams of the compound. In other words, one hundred times the total mass of all the atoms of that element divided by the total mass of all atoms of all the elements in the formula unit.

☐ Make sure to read the Examples and work the Practice Problems in the text as you read this section. Check your answers carefully.

☐ Work the Problems for this section at the end of the chapter.

Section 7.3: The Mole

The mole is a very useful measure in chemistry. It enables chemists to actually count atoms and molecules by weighing them. This is the same way nails are sold at a hardware store. By knowing how much a single nail weighs, a thousand nails can be "counted" by weighing out one thousand times the weight of a single nail.

A mole is the number of atoms that weigh the atomic mass in grams and the number of molecules, or formula units, that weigh the molecular or formula mass in grams. Look at the periodic table. The average sodium atom weighs 22.9898 amu. A mole of sodium atoms weighs 22.9898 grams. A formula unit of NaOH weighs 39.9972 amu (22.9898 + 15.9994 + 1.0080). A mole of NaOH formula units weighs 39.9972 grams. A **molecule** of carbon dioxide weighs 44.0098 *amu* (12.011 + 2 x 15.9994). A **mole** of carbon dioxide molecules weighs 44.0098 *grams*.

The molar mass of a substance is the mass that one mole of that substance weighs. For example, the molar mass of sodium is 22.9898 grams, the molar mass of sodium hydroxide is 39.9972 grams, and the molar mass of carbon dioxide is 44.0098 grams.

☐ You should memorize **Avogadro's number, 6.02 x 10^{23}**. This is the number of things in one mole. You should think of the mole as a counting unit similar to a dozen. It doesn't matter what you are counting; if you have a dozen of them you have twelve. The same is true with Avogadro's number; if you have a mole of atoms then you have 6.02 x 10^{23} atoms. If you have a mole of molecules then you have 6.02 x 10^{23} molecules.

You should now be able to perform some very important calculations involving moles. Perform each of the following conversions and compare your answer to the one in parentheses.

1. moles to grams   How many grams of sodium chloride are there in 3.20 mol of sodium chloride? (187 grams)

2. grams to moles   How many moles of calcium chloride are there in 54.3 grams of calcium chloride? (0.489 mol)

3. moles to atoms   How many atoms of silver are there in 3.20 mol of silver? (1.93 x 10^{24} atoms)

4. atoms to moles   How many moles of atoms are there in 8.54×10^{24} atoms? (14.2 mol)

5. moles to molecules   How many molecules of carbon dioxide are there in 0.456 mol of carbon dioxide? (2.75×10^{23} molecules)

6. molecules to moles   How many moles of ammonia molecules are there in 1.23×10^{22} molecules of ammonia? (0.0204 mol)

7. grams to atoms   How many atoms of chlorine are there in 7.98 grams of calcium chloride? (8.65×10^{22} atoms of chlorine)

8. atoms to grams   How many grams of aluminum oxide would contain 5.63×10^{24} atoms of oxygen? (318 grams)

9. grams to molecules   How many molecules of carbon dioxide are there in 6.78 grams of carbon dioxide? (9.28×10^{22} molecules)

10. molecules to grams   How many grams of ammonia would it take to have 2.34×10^{25} molecules of ammonia? (661 grams)

☐ Make sure to read the Examples and work the Practice Problems in the text as you read this section. Check your answers carefully.

☐ Work the Problems for this section at the end of the chapter.

Section 7.4: Empirical Formulas

Before attempting this section you should know certain decimal equivalents and how to obtain ratios from them. For example, if you are told that two things are related in a ratio of 1.5 to 1, then you should be able to write this as $3/2$ to $2/2$. Multiplying both numbers by 2 you would see the whole number ratio of 3 to 2.

Some common fractions and their decimal equivalents are:

$$0.250 = 1/4 \qquad 0.125 = 1/8 \qquad 0.167 = 1/6$$
$$0.500 = 1/2 \qquad 0.375 = 3/8 \qquad 0.333 = 1/3$$
$$0.750 = 3/4 \qquad 0.625 = 5/8 \qquad 0.667 = 2/3$$
$$ \qquad 0.875 = 7/8 \qquad 0.833 = 5/6$$

If you see a ratio of 1.333 to 1, you could rewrite this as $1\,1/3$ to 1, or $4/3$ to $3/3$. Multiplying both numbers by three, you would find a whole number ratio of 4 to 3.

If you see a ratio of 1.625 to 1.75 to 1, you would multiply by 8, because the numbers can all be expressed as eighths.

$$1.625 \;=\; 1\,5/8 \;=\; 13/8 \;=\; 13/8$$
$$1.75 \;=\; 1\,3/4 \;=\; 1\,6/8 \;=\; 14/8$$
$$1 \;=\; 8/8 \;=\; 8/8 \;=\; 8/8$$

The ratio becomes $13/8$ to $14/8$ to $8/8$, or 13 to 14 to 8.

☐ Read Section 7.4.

☐ Make sure to read the Examples and work the Practice Problems in the text as you read this section. Check your answers carefully.

☐ Work the Problems for this section at the end of the chapter.

Section 7.5: Molecular Formulas

Molecular formulas tell exactly how many atoms are in a molecule. The molecular formula is not necessarily the lowest whole number ratio of the atoms in a molecule; that is the empirical formula.

☐ Read Section 7.5.

☐ Make sure to read the Examples and work the Practice Problems in the text as you read this section. Check your answers carefully.

☐ Work the Problems for this section at the end of the chapter.

Chapter 7: Finishing Up

☐ Carefully read the summary section at the end of the chapter. Do you understand each paragraph? Do you know the terms used? If not, review the section indicated at the end of the paragraph.

☐ The section titled "Items for Special Attention" gives you some pointers to help you learn the material correctly and also can keep you from making some of the more common mistakes. Read these carefully.

☐ Work the "Self-Tutorial Problems." Check each answer as you complete the problem. If you have trouble with one of these, be sure to ask your instructor for help.

☐ Look over the list of "Key Terms" at the beginning of the chapter. If you do not recognize a term or are unsure of how it was used in the chapter, go back to that section and reread it.

☐ Use your lecture notes and the text to find the important topics covered in class. Make flash cards so that you can study these carefully before the exam.

☐ Work each of the problems at the end of the chapter in your text. Be sure to check your answers. If the answer is not given for that problem, work another that is similar and check the answer. If you are incorrect, check your work and review the chapter to see if you can answer the question correctly. If you cannot, get help from your instructor.

☐ Work the sample exam questions on the next few pages.

☐ Check your answers when you have completed all of them.

☐ Make up your own exam and exchange it with a fellow student or use it in your study group.

Chapter 7: Sample Exam Questions

1. Determine the empirical formula from the following data:

    62.1% carbon, 10.3% hydrogen, 27.6% oxygen

2. Determine the empirical formula from the following data:

    48.7% carbon, 8.1% hydrogen, 43.2% oxygen

3. How many atoms of hydrogen are there in 5.27 grams of $(NH_4)_3PO_4$?

4. How many atoms of hydrogen are there in 5.27 grams of CH_3OH?

5. Calculate the formula mass for each of the following:

   (a) $NH_4C_2H_3O_2$ _____

   (b) $Fe_2(Cr_2O_7)_3$ _____

6. What is the percentage of nitrogen in $Ca(CN)_2$?

7. How many moles are there in each of the following samples?

   (a) 27.8 grams of H_2SO_4 _____

   (b) 135.8 grams of $Ca_3(PO_4)_2$ _____

8. How many oxygen atoms are there in 3.45 grams of $Na_2S_2O_3$?

9. Calculate the mass of ammonium acetate that contains 2.54×10^{25} hydrogen atoms.

10. Calculate the mass in grams of one average sodium atom.

11. Calculate the mass in grams of one formula unit of iron(III) oxide.

12. Determine an empirical formula for a compound of carbon, hydrogen, oxygen and nitrogen from the following data:

$$C = 59.4\%$$
$$H = 10.9\%$$
$$O = 15.8\%$$
$$N = 13.9\%$$

13. Determine an empirical formula for a compound of carbon, hydrogen and oxygen from the following data:

$$C = 52.2\%$$
$$H = 13.0\%$$
$$O = 34.8\%$$

14. A compound was found to contain 85.7% carbon and 14.3% hydrogen. The molar mass was established at 84.0 g/mol. What is the molecular formula for this compound?

Chapter 7: Experiment at Home

Fun with Food

Find containers that have both English and metric units. Using the conversion values in your text, check the accuracy of the manufacturer. These will usually be quite close, but it will give you a reason to have your calculator with you at the breakfast table and will give you something new to do instead of just reading the cereal boxes.

Food is divided into three main categories: fat, protein, and carbohydrate. The energy in food is measured in kilocalories. These are often called nutritional calories and are designated by writing the unit (Calorie) with a capital C. Proteins and carbohydrates contain about four kilocalories per gram. Fat contains about nine kilocalories per gram. Using the nutritional statements on a food container, calculate how many kilocalories are in a serving based on the grams of fat, protein, and carbohydrate. For example, nutritional information per serving on a can of cashews states that there are 160 Calories per 1 ounce serving. There are 5 grams of protein, 8 grams of carbohydrate, and 13 grams of fat. To find the actual kilocalories you would do the following calculation:

$$(4 \text{ kcal/g-protein})(5 \text{ g}) + (4 \text{ kcal/g-carbohydrate})(8 \text{ g}) + (9 \text{ kcal/g-fat})(13 \text{ g})$$

From this you will find that there are actually 169 kilocalories in one ounce which is close to the value of 160 Calories on the can.

Try this on other items.

Chapter 8: Chemical Reactions

Before you get started

☐ Review the meaning of a chemical formula. Remember what the subscripts tell you about the number of atoms of each element in the formula (Section 5.1).

☐ A number in front of a formula, the coefficient, tells how many formula units there are. Every atom in the formula unit is multiplied by this number.

☐ Review how to write chemical formulas knowing the charge on the cation and the anion (Figure 5.10).

☐ Remember the important polyatomic ions that you learned in Chapter 5 (Table 5.2).

☐ Remember the elements that exist as diatomic molecules (Figure 5.3).

Section 8.1: The Chemical Equation

☐ Read Section 8.1. Note the words in **bold** type. These are important to remember. You need to know the difference between a reactant (reagent) and a product.

☐ Read the following statement, fill in your name and sign on the line below:

The Equation Promise

I, _____, promise that, from this day forward, whenever I see a chemical equation, whether it be on a quiz, an exam or even in a textbook, I will carefully examine it to make sure it is balanced.

signed: _____

date: _____

☐ Make sure to read the Examples and work the Practice Problems in the text as you read this section. Check your answers carefully.

☐ Work the Problems for this section at the end of the chapter.

Section 8.2: Balancing Equations

Your book points out two very common mistakes made in balancing equations. The first is trying to change the universe to fit the equation. Many students try to put together formulas with the wrong ratios in order to make an equation balance. You learned in Chapter 5 about bonding. Elements form compounds according to rules. You cannot change the ratios of atoms in formulas to make balancing equations easier. The only number you can insert is the coefficient in front of the formula. The second is the problem of remembering that there is a *1* in front of a formula in a balanced equation even though you do not see it. The suggestion of the question mark is a good one.

When balancing equations, it is wise to start with the more complex substances. Also, when an element appears in more than one substance on either side of the equation, you should leave that element until last. The nice part about balancing equations is that you will know when you have the right answer. Just inventory the atoms on both sides of the equation carefully to see that there are the same number of each kind. If there are, then the equation is balanced.

☐ Read Section 8.2.

Evidence of a chemical reaction is the formation of heat, a solid (precipitate), or a gas. These are generally indicated by symbols which will become familiar to you in this chapter. Heat is sometimes shown by the Greek letter delta, Δ. Precipitates (insoluble solids) are designated by placing (s) after the compound. For example, the precipitate of silver chloride is shown as AgCl(s). Gases present in or generated by a reaction are shown by placing (g) after the compound. Carbon dioxide would appear as CO_2(g). When molecules such as water are present as a liquid in a reaction (l) is used. For example, water would be H_2O(l). Be careful; many reactions are run in water solution. Only when water is a reactant or a product is it shown in the equation. Substances that are dissolved in water are designated by (aq).

☐ Make sure to read the Examples and work the Practice Problems in the text as you read this section. Check your answers carefully.

☐ Work the Problems for this section at the end of the chapter.

Section 8.3: Predicting the Products of Chemical Reactions

There are many types of reactions, but most fall into the five categories explained in this section. As you read about each type of reaction, think about two things: How do I recognize this type of reaction? How do I know what the products are going to be?

☐ Read Section 8.3 at this time.

☐ After each name below, write an equation for an example of the reaction. Be sure to pay special attention to balancing each equation.

1. combination reaction: _____

2. decomposition reaction: _____

3. substitution reaction: _____

4. double substitution reaction: _____

5. combustion reaction: _____

Combination and **decomposition** reactions are fairly easy to understand. Be careful when dealing with diatomic elements. When the element is uncombined, the molecules are diatomic. Also note that noble gases do not readily react.

Substitution reactions involve one element replacing another in a compound. Reactivity as shown on Table 8.2, can best be explained as an element's tendency to want to combine to form a compound. Sodium metal, which is found at the top of the table, is never found in nature in the elemental form. It is always combined with other elements to form compounds. The coinage metals are found near the bottom of the table. These metals do not form compounds easily and tend to keep their luster and metallic character.

You must follow the steps **in order** to properly write **double substitution** equations.

1. Identify the cation and anion of each compound present on the reactant side of the reaction. Be careful to identify any polyatomic ions that are present.
2. Write formulas for the products by exchanging the partners. Use the rules of compound formation from Chapter 5.
3. Balance the equation.

Any soluble ionic compound will dissociate into ions. Tables 5.2, 6.2, 6.4, and 6.5 show you many of the polyatomic ions. When a substance is insoluble, according to the rules of Table 8.3, the ions do not separate, but form a precipitate. If H_2CO_3 or NH_4OH would be formed, follow the rules as shown in Example 8.9.

Combustion reactions are reactions with oxygen.

☐ Make sure to read the Examples and work the Practice Problems in the text as you read this section. Check your answers carefully.

☐ Work the Problems for this section at the end of the chapter.

Section 8.4: Acids and Bases

Reactions involving acids and bases are commonly described as proton transfer reactions. In this case, the proton is the hydrogen ion, H^+. Acids give up H^+ and bases accept H^+. When a substance is an acid and has an ionizable hydrogen atom, the formula for the compound is written so that the H comes first in the formula.

☐ Read Section 8.4.

When an acid reacts with a base, a *neutralization* reaction occurs. The products of this reaction are a salt and water. When there are the same number of hydrogen ions as there are hydroxide ions, the solution is neutral. To show the approximate H^+ concentration of a solution, an *indicator* is often used. These substances changes color depending on whether they are in an excess of hydrogen ions or an excess of hydroxide ions.

Acid and base strength is measured by how much the molecules dissociate in water. A strong acid will completely dissociate in water. A weak acid will only partially dissociate in water and the concentration of H^+ ions will be a fraction of the concentration of the acid. You should memorize the strong acids and bases at this time. Any acid or base that is not on the list is considered a weak acid or base. Be sure to know these before you study Chapter 17.

Strong acids	HCl HBr HI $HClO_3$ $HClO_4$ HNO_3 H_2SO_4
Strong bases	LiOH NaOH KOH RbOH CsOH $Ba(OH)_2$

The reaction of an acid with a metal above hydrogen on the activity series of Table 8.2 will result in the formation of a salt and hydrogen gas. Note that this is diatomic hydrogen gas.

Acidic and basic anhydrides are formed by the removal of water from an acid or a base. When water is added to an acidic anhydride, an acid is formed. When water is added to a basic anhydride, a base is formed.

Your textbook mentions the reaction of SO_2 with the water in the air to form sulfurous acid. There is also a reaction of NO_2 with the water in the air to form nitric acid. The formation of the NO_2 occurs as a result of lightning causing the N_2 in the air to combine with O_2. Plants are unable to break the triple bond of diatomic nitrogen molecules to form the nitrogen needed for the formation of amino acids and other important biochemicals. This form of acid rain is therefore very beneficial because it serves as a primary source of nitrogen for plant nutrition.

☐ Read carefully the equations in this section. Are they all balanced? Can you see how the reactants change partners? When water is one of the products, it is easier to see the exchange if you write water as HOH.

$$HCl(aq) + NaOH(aq) \rightarrow NaCl(aq) + HOH(l)$$

When only one hydrogen ion is removed from a polyprotic acid (an acid with more than one ionizable hydrogen), the result is an acid salt. Acid salts of phosphoric and carbonic acid are responsible for maintaining the acid balance in blood. You will learn more about this in Chapter 17. Figure 8.11 shows the reactions of carbonates and acid carbonates. Be sure you can follow the diagrams of this table.

☐ Make sure to read the Examples and work the Practice Problems in the text as you read this section. Check your answers carefully.

☐ Work the Problems for this section at the end of the chapter.

Chapter 8: Finishing Up

☐ Carefully read the summary section at the end of the chapter. Do you understand each paragraph? Do you know the terms used? If not, review the section indicated at the end of the paragraph.

☐ The section titled "Items for Special Attention" gives you some pointers to help you learn the material correctly and also can keep you from making some of the more common mistakes. Read these carefully.

☐ Work the "Self-Tutorial Problems." Check each answer as you complete the problem. If you have trouble with one of these, be sure to ask your instructor for help.

☐ Look over the list of "Key Terms" at the beginning of the chapter. If you do not recognize a term or are unsure of how it was used in the chapter, go back to that section and reread it.

☐ Use your lecture notes and the text to find the important topics covered in class. Make flash cards so that you can study these carefully before the exam.

☐ Work each of the problems at the end of the chapter in your text. Be sure to check your answers. If the answer is not given for that problem, work another that is similar and check the answer. If you are incorrect, check your work and review the chapter to see if you can answer the question correctly. If you cannot, get help from your instructor.

☐ Work the sample exam questions on the next few pages.

☐ Check your answers when you have completed all of them.

☐ Make up your own exam and exchange it with a fellow student or use it in your study group.

Chapter 8: Sample Exam Questions

1. Balance the equation:

$$C_6H_{14} + O_2 \rightarrow CO_2 + H_2O$$

2. Balance the equation:

$$C_7H_{16} + O_2 \rightarrow CO_2 + H_2O$$

3. Write a balanced equation for the following:

   Aqueous sodium hydroxide and aqueous phosphoric acid react to produce aqueous sodium phosphate and water.

4. Write a balanced equation for the following:

   Aqueous barium hydroxide and aqueous phosphoric acid react to produce solid barium phosphate and water.

5. Identify the type of reaction and predict the products for the following:

$$Zn(s) + Cu(NO_3)_2 \rightarrow$$

   This is an example of a _____ reaction.

$$Zn(s) + Cu(NO_3)_2 \rightarrow \underline{\qquad} + \underline{\qquad}$$

6. Identify the type of reaction and predict the product for the following:
    (Do not forget to balance the equation!)

$$Mg(s) + O_2 \rightarrow$$

This is an example of a _____ reaction.

$$Mg(s) + O_2 \rightarrow \text{_____}$$

7. Balance the following equations:

    (a) __ Al + __ NaOH + __ H_2O → __ $NaAlO_2$ + __ H_2

    (b) __ C + __ Fe_2O_3 → __ CO + __ Fe

8. Write balanced equations for the following reactions:

    (a) Sulfur and oxygen combine to form sulfur dioxide.

    (b) Zinc replaces copper in copper(II) chloride.

    (c) Tin(II) chloride combines with chlorine to produce tin(IV) chloride.

9. Write balanced equations for the following reactions:

    (a) Ammonium carbonate is mixed with barium nitrate.

    (b) Potassium chlorate decomposes to form oxygen and potassium chloride.

    (c) Silver combines with sulfur.

10. Complete and balance the following equations.

   (a) $HCl + NaOH \rightarrow$

   (b) $Na_2CO_3 + HCl \rightarrow$

   (c) $Pb(NO_3)_2 + NaCl \rightarrow$

11. Complete and balance the following equations.

   (a) $H_2 + O_2 \rightarrow$

   (b) $Mg + HCl \rightarrow$

   (c) $Na_2S + Ni(NO_3)_2 \rightarrow$

12. Tell the reaction type and name the products formed. Be sure to write a balanced equation.

   (a) $C_6H_{14} + O_2(excess) \rightarrow$

   Reaction type: _____

   Products formed: _____

   (b) $Ba(OH)_2 + H_2SO_4 \rightarrow$

   Reaction type: _____

   Products formed: _____

   (c) $SO_3 + CaO \rightarrow$

   Reaction type: _____

   Products formed: _____

13. Tell the reaction type and name the products formed. Be sure to write a balanced equation.

   (a) NH_4HCO_3 + heat →

   Reaction type: _____

   Products formed: _____

   (b) $AlCl_3$ + Na_2CO_3 →

   Reaction type: _____

   Products formed: _____

   (c) Al + O_2 →

   Reaction type: _____

   Products formed: _____

14. Using the solubility rules, predict whether a precipitate will form in the following double substitution reactions. If so, complete and balance the equation. Write the formula for the precipitate on the line. If no reaction occurs, write NR.

   (a) $Pb(NO_3)_2$ and $NaC_2H_3O_2$ _____

   (b) $BaCl_2$ and Na_3PO_4 _____

   (c) $(NH_4)_2SO_4$ and $Al(NO_3)_3$ _____

   (d) $Ca(NO_3)_2$ and $NH_4C_2H_3O_2$ _____

Chapter 8: Experiment at Home

Acid and Base Test Solutions

Sometimes it is very important to know if the solution you have is an acid or a base. In this experiment you will be given formulas for two test solutions. You will then be able to use these solutions to determine whether other solutions have an excess of hydrogen ions (acidic) or an excess of hydroxide ions (basic).

Materials: baking soda
two laxative tablets, containing phenolphthalein
two jars with lids for storing solutions
two eyedroppers, one for each jar
test tubes, jars, cups or small glasses
household solutions to test

Acid Test Solution

Carbonates are compounds containing the carbonate ion (CO_3^{2-}). Bicarbonates are compounds containing the bicarbonate ion (HCO_3^-). When either carbonates or bicarbonates react with excess acid, carbon dioxide and water are given off. The effervescence of a carbonate or bicarbonate solution indicates the presence of hydrogen ions.

Add about one teaspoon of baking soda ($NaHCO_3$) to one half cup of water. Store this acid test solution in a small jar. To test for an acid, place a small amount of the unknown solution in one of your test containers. Add a few drops of acid test solution. If the combination effervesces, you can conclude that the unknown solution is acidic.

Some common household acids that you might wish to try include vinegar, orange juice, and vitamin C tablets.

Base Test Solution

An indicator is a dye that has a visible difference between its form in an acidic solution and its form in a basic solution. Phenolphthalein is an acid-base indicator. It is colorless in acidic solution but turns bright pink in basic solution.

With a spoon crush two laxative tablets containing phenolphthalein and place them in a jar. Add sufficient water to dissolve them to make your base test solution. To test for a base, place a small amount of the solution to be tested in one of your test containers and add a few drops of your base test solution. If the combination turns pink, you can conclude that the substance you tested is a base.

Common bases you may wish to test include ammonia and a solution of laundry detergent.

Chapter 9: Net Ionic Equations

Before you get started

☐ Review the meaning of a chemical formula. Remember what the subscripts tell you about the number of atoms of each type in the formula (Section 5.1).

☐ The number in front of a formula, the coefficient, tells how many formula units there are. Every atom in the formula unit is multiplied by this number.

☐ Review how to write chemical formulas knowing the charges on the cation and the anion (Figure 5.10).

☐ Remember the important polyatomic ions that you learned in Chapter 5 (Table 5.2).

☐ Remember the elements that exist as diatomic molecules (Figure 5.3).

☐ Remember how to write a balanced chemical equation (Chapter 8).

☐ Remember the types of chemical reactions and the products of certain reactions. (Chapter 8).

A net ionic equation results when all unreacting ions are removed from the equation. These ions will still be present in the solution, but they are there only to balance the charge. The overall charge on a solution will always be neutral. The net ionic equation simply describes the actual reaction taking place.

Section 9.1: Properties of Ionic Compounds in Aqueous Solution

☐ Read Section 9.1.

When working the Examples and Practice Problems of this chapter, it is important to remember that soluble counterions are present in the solution in order to maintain neutrality. These are very often the sodium ion (Na^+), the potassium ion (K^+), the ammonium ion (NH_4^+), the nitrate ion (NO_3^-), and the acetate ion ($C_2H_3O_2^-$).

You may wish to make a copy of Table 8.3 to use when deciding if a compound is soluble. Substances which ionize or dissociate completely in aqueous solution are listed in this section. You should memorize these and the strong acids and bases in Table 9.1.

Strong Acids and Bases

Strong acids	HCl HBr HI $HClO_3$ $HClO_4$ HNO_3 H_2SO_4
Strong bases	LiOH NaOH KOH RbOH CsOH $Ba(OH)_2$

Net ionic equations are used to explain the reaction in simplified terms. Only those species reacting are included in a net ionic equation. In exchange reactions, there are three products that are commonly formed. You will need to be able to identify these before you can write net ionic equations.

1. A precipitate — When a combination of ions results in a compound that is insoluble in water.
2. A gas — When a combination of ions results in the formation of a gaseous product.
3. Water — When water is formed as a product of the reaction.

☐ Make sure to read the Examples and work the Practice Problems in the text as you read this section. Check your answers carefully.

☐ Work the Problems for this section at the end of the chapter.

Section 9.2: Writing Net Ionic Equations

☐ Read Section 9.2.

Look at Figure 9.4. It is important to remember that solutions are always neutral. That is, the charges present on the ions in the solution are always balanced. There is no such thing as a bottle of chloride ions. There is always some *counterion* present to balance the charge.

☐ Make sure to read the Examples and work the Practice Problems in the text as you read this section. Check your answers carefully.

☐ Work the Problems for this section at the end of the chapter.

Chapter 9: Finishing Up

☐ Carefully read the summary section at the end of the chapter. Do you understand each paragraph? Do you know the terms used? If not, review the section indicated at the end of the paragraph.

☐ The section titled "Items for Special Attention" gives you some pointers to help you learn the material correctly and also can keep you from making some of the more common mistakes. Read these carefully.

☐ Work the "Self-Tutorial Problems." Check each answer as you complete the problem. If you have trouble with one of these, be sure to ask your instructor for help.

☐ Look over the list of "Key Terms" at the beginning of the chapter. If you do not recognize a term or are unsure of how it was used in the chapter, go back to that section and reread it.

☐ Use your lecture notes and the text to find the important topics covered in class. Make flash cards so that you can study these carefully before the exam.

☐ Work each of the problems at the end of the chapter in your text. Be sure to check your answers. If the answer is not given for that problem, work another that is similar and check the answer. If you are incorrect, check your work and review the chapter to see if you can answer the question correctly. If you cannot, get help from your instructor.

☐ Work the sample exam questions on the next few pages.

☐ Check your answers when you have completed all of them.

☐ Make up your own exam and exchange it with a fellow student or use it in your study group.

Chapter 9: Sample Exam Questions

1. Write the net ionic equation for the reaction of sodium carbonate and excess hydrochloric acid.

2. Write the net ionic equation for the reaction of aqueous barium hydroxide and sulfuric acid.

3. Write the net ionic equation for the reaction of sodium hydrogen carbonate and excess hydrochloric acid.

4. Predict whether each of the following is soluble:

    (a) AgI _____

    (b) $(NH_4)_2CO_3$ _____

    (c) $Zn_3(PO_4)_2$ _____

    (d) K_2S _____

    (e) Al_2O_3 _____

5. Complete and balance an equation for each of the following reactions. Then write the net ionic equation. Be sure to indicate if there is no reaction, and thus no net ionic equation.

(a) $CaCl_2(aq) + K_3PO_4(aq)$

(b) $FeCl_2(aq) + NaOH(aq)$

(c) $NaHCO_3(aq) + H_2SO_4(aq)$

(d) $Na_2S(aq) + KCl(aq)$

(e) $Ba(OH)_2(aq) + HCl(aq)$

6. Many salts containing heavy metal ions such as silver (Ag^+), lead (Pb^{2+}), and mercury (Hg^+) are toxic if taken internally. What anion could be used to make a test solution that would cause a precipitate to form in the presence of these cations?

Write chemical equations for the following reactions. If there is a reaction, write the net ionic equation. If no reaction occurs, write NR.

7. Potassium hydroxide completely neutralizes sulfuric acid.

8. Iron(III) nitrate and sodium hydroxide solutions are combined.

9. Zinc is placed into a solution of hydrochloric acid.

10. Hydrochloric acid is added to a solution of sodium hydrogen carbonate.

11. Solutions of silver nitrate and sodium chloride are combined.

Chapter 9: Experiment at Home

Chromatography

Chromatography is used in many laboratories to separate the components of mixtures. Forensic laboratories detect the drugs present in a sample, hospital labs can tell the relative concentrations of various enzymes, and many municipal water supplies are checked for trace contaminants using chromatographic methods.

Materials One screw-capped, wide-mouthed glass jar with lid (a pint canning jar works well)
heavy typing paper (25% cotton)
pencil
scissors
stapler
alcohol
vinegar
ammonia
various markers and pens

Method: Make a cylinder that will stand upright in the glass jar from a strip of heavy typing paper. The paper must be long enough to fit standing on end inside the jar and wide enough to wrap around and staple. With a pencil, draw a line one inch from the bottom of the paper (the longer side) along the entire length of the paper. Make small spots with the different markers and pens at about one inch intervals. While the spots are drying, place about ½ inch of alcohol in the glass jar. Place the lid on the jar and allow the vapors to fill the jar. Gently put the paper into the jar with the spots on the bottom side. Make sure the paper is in the alcohol, but the alcohol should not be deep enough to touch the spots. As the alcohol is drawn up the paper, the spots will be separated into components based upon the components' solubility in the alcohol. When the alcohol has travelled up the paper to about one inch from the top, remove the paper and allow it to dry.

Theory: The distance a substance travels on the paper (stationary phase) is dependent upon its solubility in the mobile phase (solvent). The solubility of compounds is dependent upon the nature of its molecules. If the molecules are polar, then the substance will be more soluble in polar solvents such as water and alcohol. In addition, the solubility of a compound can be altered by making the solvent more acidic (adding vinegar) or more basic (adding ammonia).

Additional Projects: (1) Try placing dots from several markers together and see if they can be separated. (2) Repeat the experiment using alcohol with vinegar added. (3) Repeat the experiment using alcohol with ammonia added. (4) Try the experiment on various plant extracts. (Coffee, tea, tomato paste, *etc.*)

Chapter 10: Stoichiometry

Before you get started

☐ Review the meaning of a chemical formula. Remember what the subscripts tell you about the number of atoms of each type in the formula (Section 5.1).

☐ The number in front of a formula, the coefficient, tells how many formula units there are. Every atom in the formula unit is multiplied by this number.

☐ Review how to write chemical formulas knowing the charges on the cation and the anion (Figure 5.10).

☐ Remember the important polyatomic ions that you learned in Chapter 5 (Table 5.2).

☐ Remember the elements that exist as diatomic molecules (Figure 5.3).

☐ Remember how to calculate the molar mass of a compound. (Chapter 7)

☐ Remember how to find the number of moles of a substance if you know the number of grams and how to find the number of grams if you know the number of moles. (Chapter 7)

One of the most important calculations a chemist performs is predicting the amount of product from a given amount of reactants. In this chapter you will use your knowledge of balanced equations to do this.

Section 10.1: Mole Calculations for Chemical Reactions

☐ Read Section 10.1.

Whenever you read a chemical equation remember that you can use the word "mole" instead of atom or molecule or formula unit. The ratio of the coefficients of any two substances in a balanced chemical equation is called the **mole ratio** or **reacting ratio.** As you can see from Example 10.1, any two substances in an equation can be related in a ratio form. Example 10.2 shows you how you will be using these ratios to perform calculations.

Take a moment and look back at Chapter 2. You learned how to use conversion factors to change from one unit to another. Here you are converting from moles of one substance to moles of another in a balanced chemical equation. The same principles apply to mole ratios that applied for other conversion factors.

☐ Make sure to read the Examples and work the Practice Problems in the text as you read this section. Check your answers carefully.

☐ Work the Problems for this section at the end of the chapter.

Section 10.2: Mass Calculations for Chemical Reactions

In Chapter 7 you learned how to convert from moles to grams and from grams to moles. In this section you will use what you learned in Section 10.1 and Chapter 7 to start with grams of a substance and predict how many grams of product will be produced.

☐ Read Section 10.2.

Establishing a *unit path* through a problem is very helpful, especially when there are several steps in the problem. The diagram in Figure 10.2 shows you this.

                      ①   ②   ③

grams of A   →   moles of A   →   moles of B   →   grams of B

Each of the arrows corresponds to a conversion factor. A and B are substances in a balanced chemical equation. ① is the factor from the molar mass of A. ② is the mole ratio from the balanced chemical equation. ③ is the molar mass of B.

☐ Make sure to read the Examples and work the Practice Problems in the text as you read this section. Check your answers carefully.

☐ Work the Problems for this section at the end of the chapter.

Section 10.3: Calculations Involving Other Quantities

☐ Read Section 10.3.

☐ Carefully look at the chart in Figure 10.3. The boxes give the quantities and the arrows show the conversion factors necessary to go between the two. For example, to convert from volume of reactant A to mass of reactant A, you would multiply by the density. If you wanted to go from the mass of reactant A to the volume of reactant A, you would divide by the density. Use this figure to write some sample questions on stoichiometry. Construct problems which contain as many of the factors as possible. Compare your problems with those of the other members of your study group.

☐ Make sure to read the Examples and work the Practice Problems in the text as you read this section. Check your answers carefully.

☐ Work the Problems for this section at the end of the chapter.

Section 10.4: Problems Involving Limiting Quantities

You have seen several problems where one of the reactants was said to be in *excess*. This was done so that you would know which reactant to use in the calculation. The reactant that is used up first is the one which will determine how much product can be formed. It is called the *limiting reactant* because it limits the amount of product. Some of the reactant in excess will be left over when the reaction is over.

This type of problem is easy to recognize since you will always be given the quantity of at least two reactants. From what you learned in the previous sections of this chapter, you know that you only need to know the amount of *one* of the reactants or products to tell how much of any other reactant or product is present.

☐ Read Section 10.4. Remember that it is imperative to have a balanced equation with which to form the mole ratios.

☐ Look over the chart in Figure 10.4.

Although it requires more calculations (and more chances for error), many students find that by simply calculating the amount of product using each reactant quantity, the limiting reactant can be identified by finding the one that produces the **least** amount of product. Looking at Example 10.17, you can calculate the amount of product formed by the given amount of each reactant.

$$2.90 \text{ mol } Cl_2 \times \frac{2 \text{ mol NaCl}}{1 \text{ mol } Cl_2} = 5.80 \text{ mol NaCl}$$

$$4.90 \text{ mol Na} \times \frac{2 \text{ mol NaCl}}{2 \text{ mol Na}} = 4.90 \text{ mol NaCl}$$

Since 4.90 mol of NaCl is less than 5.80 mol of NaCl, you can easily see that the 4.90 mol of Na is the limiting reactant and that the 2.90 mol of Cl_2 is the reactant in excess. The reactant that is the limiting reactant will be used up first. This method becomes more complicated as the number of reactants and products increases. The method in your book is preferred for the calculation; the method above is given to help you understand the concept.

☐ Make sure to read the Examples and work the Practice Problems in the text as you read this section. Check your answers carefully.

☐ Work the Problems for this section at the end of the chapter.

Section 10.5: Theoretical Yield and Percent Yield

☐ Read Section 10.5.

When working with percent yield, remember that the actual yield will always be less than or equal to the theoretical yield.

☐ Make sure to read the Examples and work the Practice Problems in the text as you read this section. Check your answers carefully.

☐ Work the Problems for this section at the end of the chapter.

Section 10.6: Calculations with Net Ionic Equations

☐ Read Section 10.6.

When you are working problems with net ionic equations, as with all stoichiometry problems, remember that you must be working with a **balanced** equation.

☐ Make sure to read the Examples and work the Practice Problems in the text as you read this section. Check your answers carefully.

☐ Work the Problems for this section at the end of the chapter.

Chapter 10: Finishing Up

☐ Carefully read the summary section at the end of the chapter. Do you understand each paragraph? Do you know the terms used? If not, review the section indicated at the end of the paragraph.

☐ The section titled "Items for Special Attention" gives you some pointers to help you learn the material correctly and also can keep you from making some of the more common mistakes. Read these carefully.

☐ Work the "Self-Tutorial Problems." Check each answer as you complete the problem. If you have trouble with one of these, be sure to ask your instructor for help.

☐ Look over the list of "Key Terms" at the beginning of the chapter. If you do not recognize a term or are unsure of how it was used in the chapter, go back to that section and reread it.

☐ Use your lecture notes and the text to find the important topics covered in class. Make flash cards so that you can study these carefully before the exam.

☐ Work each of the problems at the end of the chapter in your text. Be sure to check your answers. If the answer is not given for that problem, work another that is similar and check the answer. If you are incorrect, check your work and review the chapter to see if you can answer the question correctly. If you cannot, get help from your instructor.

☐ Work the sample exam questions on the next few pages.

☐ Check your answers when you have completed all of them.

☐ Make up your own exam and exchange it with a fellow student or use it in your study group.

Chapter 10: Sample Exam Questions

1. Calculate the mass of silver chloride that can be produced from the reaction of 125.0 grams of silver nitrate and 50.00 grams of sodium chloride.

2. After a 10.00-gram sample of $KClO_3$ was heated for five minutes, 6.22 grams of powder was left. The powder contained KCl. $KClO_3$ decomposes into O_2 and KCl. Was the reaction complete? If not, how much KCl should be left when all of the oxygen is removed?

3. Calculate the mass of barium carbonate that can be produced from the reaction of 125.0 grams of barium nitrate and 50.00 grams of sodium carbonate.

4. Magnesium reacts with oxygen to produce magnesium oxide. After ten minutes of heating in an oxygen atmosphere, the product of a 10.00-gram sample of magnesium weighs 15.38 grams. Has all of the magnesium reacted? If not, how much magnesium oxide will there be when the reaction is complete?

5. How many grams of hydrogen gas is produced when 24.8 grams of zinc is treated with phosphoric acid?

6. Aluminum is burned in air to produce aluminum oxide. How many grams of oxygen is needed to burn 249.6 grams of aluminum?

7. Beginning with 9.00 moles of Al, 8.00 moles of NaOH, and 6.00 moles of H_2O, how many grams of $NaAlO_2$ can be formed? [Don't forget to balance the equation!]

$$Al + NaOH + H_2O \rightarrow NaAlO_2 + H_2$$

8. Calculate the number of oxygen atoms in 275 mL of water.
    (Assume the density of water is 1.00 g/mL.)

9. Ethyl alcohol, C_2H_6O, is a product of the fermentation of a sugar, $C_6H_{12}O_6$. How many liters of alcohol result from the fermentation of 10.0 kilograms of sugar? (The density of alcohol is 0.789 g/mL) The equation for this reaction is:

$$C_6H_{12}O_6 \rightarrow 2\,C_2H_6O + 2\,CO_2$$

10. How many grams of CO_2 can be produced from the complete combustion of 25.0 liters of CH_4 of density 0.714 g/L and 25.0 liters of oxygen of density 1.43 g/L?

11. If 40.00 grams of sodium sulfate was treated with 25.00 grams of barium chloride, and 21.85 grams of barium sulfate was recovered, what is the percentage yield of the reaction?

12. Batrachotoxin, $C_{31}H_{42}N_2O_6$, is an arrow poison extracted from the skin of the Columbian arrow poison frog. A lethal dose is 0.05 µg. How many molecules are in a lethal dose?

13. Aniline is an organic compound of carbon, hydrogen, and nitrogen. When a 3.00 milligram sample is burned, 8.52 mg of CO_2 and 2.03 mg of H_2O are formed. What is the empirical formula for aniline? (Hint: The mass of carbon in the aniline is the same as the mass of carbon in the CO_2, the mass of hydrogen is the mass of hydrogen in the water. The mass of nitrogen is the rest of the 3.00 mg.)

14. Carbon dioxide is removed from air by treating it with sodium hydroxide to produce sodium carbonate and water. It is estimated that over a 24-hour period a person exhales about 1.0 kg of CO_2. How many kilograms of sodium hydroxide is required to remove the carbon dioxide that is formed in a 14-day shuttle mission involving six astronauts?

15. The net ionic equation for the Breathalyzer test used to determine alcohol (C_2H_6O) concentration in the body is:

$$16\ H^+ + 2\ Cr_2O_7^{2-} + 3\ C_2H_6O \rightarrow 3\ C_2H_4O_2 + 4\ Cr^{3+} + 11\ H_2O$$

How many grams of $K_2Cr_2O_7$ must be used to consume 10.0 mg of alcohol?

16. If you want to precipitate 15.98 mg of barium ions out of a solution as insoluble barium sufate, what is the minimum number of grams of sodium sulfate that you would add?

Chapter 10: Experiment at Home

Baking A Cake

Many recipes contain a large number of ingredients. Look through a cookbook and find a recipe for a cake. Divide the amounts in the recipe by 6. Design an experiment to find out what will happen if just one of the ingredients is omitted. For each of your trials use four cupcake wells each lined with a paper cupcake holder.

1) For a control, prepare one-sixth of the recipe exactly as written. These cupcakes should appear normal.

2) For another one-sixth of the recipe omit the sugar.

3) For another one-sixth omit the baking powder.

4) Omit the eggs in the next group.

5) Leave out the oil in one group.

6) Finally, for the last batch, double the amount of sugar.

Baking a cake is a chemical reaction. It is very complex, but as you will see if there is an insufficient amount of any of the reactants, you will not obtain the desired product.

Chapter 11: Molarity

Before you get started

☐ Remember how to calculate the number of moles if you know the number of grams of a substance. (Chapter 7)

☐ Remember how to calculate the number of grams if you know the number of moles of a substance. (Chapter 7)

☐ Remember that many compounds are ionic and separate into ions in solution. You should also remember that you may only see some of these ions in a net ionic equation. (Chapter 9)

☐ Remember the methods you learned in Chapter 10 about stoichiometry.

Section 11.1: Definition and Uses of Molarity

In all of chemistry, the study of how elements and compounds react with one another is based on the fact that ions, molecules, and atoms react. Since the mole has enabled chemists to measure an exact number of ions, molecules, and atoms, it is logical that concentration units for solutions should also be based on the mole.

☐ Read carefully the definitions of solute and solvent. Then read Section 11.1.

Molarity is defined as *moles* of solute per *liter* of solution. If some other unit is given, you must convert it to moles and liters in order to calculate the molarity. Molarity can also be expressed as *millimoles* of solute per *milliliter* of solution.

The term *millimole* is often used in biochemistry. Many compounds in the body are measured in millimolar amounts. Millimoles of solute per milliliter of solution is the same value as moles of solute per liter of solution.

Note that in Practice Problem 11.5 molarity is used as a conversion factor. If you are having trouble using molarity refer back to Chapter 2 where conversion factors are introduced.

The concentration of a substance decreases as the solution is diluted. If 5 mL of a 2 M solution is diluted to 50 mL (or diluted by a factor of 10) the solution concentration is divided by 10, 2 M ÷ 10, or 0.2 M.

Figure 11.2 gives the interrelationships of the terms of this chapter. See if you can make up problems for each of the paths of this figure. Try to find problems where as many paths as possible are included. Refer back to Chapter 10 if necessary.

Example 11.11 reintroduces the concept of limiting reactant. Refer back to Section 10.4 if you are having trouble.

☐ Make sure to read the Examples and work the Practice Problems in the text as you read this section. Check your answers carefully.

☐ Work the Problems for this section at the end of the chapter.

Section 11.2: Molarities of Ions

The concentration of ions is of importance in biochemistry and medicine. Often the ions, which are called *electrolytes*, found in the body are measured in molar quantities. The concentrations of sodium, potassium, and calcium ions are commonly measured by medical technologists to give a doctor an indication of any ion imbalance in a patient's body.

As you read this section, remember what you learned about how compounds dissolve. Recall that polyatomic ions such as PO_4^{3-}, SO_4^{2-}, NO_3^-, and NH_4^+ do not come apart into separate atoms or ions.

☐ Read Section 11.2.

☐ Make sure to read the Examples and work the Practice Problems in the text as you read this section. Check your answers carefully.

☐ Work the Problems for this section at the end of the chapter.

Section 11.3: Titration

Titration is one of the most widely used analytical techniques available to chemists. There are very few chemistry laboratories that do not use titration in some form for the analysis of samples. Although not limited to acids and bases, the concept of titration is most easily understood using acids and bases. When the hydrogen ions are in excess, the solution is acidic. Any indicator added to the solution will have the color of the acid form of the indicator. When the hydroxide ions are in excess, the solution is basic. Any indicator added to the solution will have the color of the base form of the indicator. When the solution changes color, the *end point* of the titration has been established. This should be very close to the point where the number of hydrogen ions in solution are equal to the number of hydroxide ions in solution. This is called the *equivalence point*. Titration is performed using many different materials for indicators and also may be accomplished by measuring the conductivity or the potential of a solution.

☐ Read Section 11.3.

The terminology associated with titration is seen in the boldface print of this section. Give a definition of each of the following terms in your own words:

titration: _____

pipet: _____

buret: _____

end point: _____

indicator: _____

Some other terms you may hear associated with titration which do not appear in your book are:

    equivalence point: The same number of moles of charges. The endpoint is the point in the titration where the indicator changes. Theoretically the endpoint should be as close to the equivalence point as possible.

    equimolar: The same number of moles.

☐ Make sure to read the Examples and work the Practice Problems in the text as you read this section. Check your answers carefully.

☐ Work the Problems for this section at the end of the chapter.

Chapter 11: Finishing Up

☐ Carefully read the summary section at the end of the chapter. Do you understand each paragraph? Do you know the terms used? If not, review the section indicated at the end of the paragraph.

☐ The section titled "Items for Special Attention" gives you some pointers to help you learn the material correctly and also can keep you from making some of the more common mistakes. Read these carefully.

☐ Work the "Self-Tutorial Problems." Check each answer as you complete the problem. If you have trouble with one of these, be sure to ask your instructor for help.

☐ Look over the list of "Key Terms" at the beginning of the chapter. If you do not recognize a term or are unsure of how it was used in the chapter, go back to that section and reread it.

☐ Use your lecture notes and the text to find the important topics covered in class. Make flash cards so that you can study these carefully before the exam.

☐ Work each of the problems at the end of the chapter in your text. Be sure to check your answers. If the answer is not given for that problem, work another that is similar and check the answer. If you are incorrect, check your work and review the chapter to see if you can answer the question correctly. If you cannot, get help from your instructor.

☐ Work the sample exam questions on the next few pages.

☐ Check your answers when you have completed all of them.

☐ Make up your own exam and exchange it with a fellow student or use it in your study group.

Chapter 11: Sample Exam Questions

1. Calculate the concentration of each ion in solution after 276 mL of 1.345 M $NaNO_3$ is mixed with 378 mL of 1.478 M $Mg(NO_3)_2$ and then diluted to 2.00 L.

2. What volume of 0.0876 M NaOH would be required to completely neutralize 25.00 mL of 0.148 M H_2SO_4?

3. What is the molarity of a solution prepared by diluting 25.00 mL of 0.6783 M HCl to 90.00 mL?

4. Calculate the concentration of each ion in solution after 173 mL of 0.5612 M NaCl is mixed with 932 mL of 1.123 M $MgCl_2$ and then diluted to 3.00 L.

5. What volume of 0.267 M NaOH is required to completely neutralize 25.00 mL of 0.175 M H_2SO_4?

6. What is the molarity of a solution prepared by diluting 5.00 mL of 0.8163 M HCl to 30.00 mL?

7. Calculate the molarity of a solution containing 45.6 grams of sodium chloride in 5.00 liters of water.

8. How many grams of calcium chloride are there in 478 mL of a 0.789 M solution?

9. How many milliliters of 1.35 M sodium hydroxide will contain 5.38 grams of sodium hydroxide?

10. What is the molarity of a solution prepared by diluting 50.00 mL of 0.1234 M HNO_3 to 150.0 mL?

11. Calculate the concentration of each ion in 1.45 M $(NH_4)_3PO_4$.

12. Calculate the concentration of each ion in solution after 25.00 mL of 0.123 M NaCl is mixed with 25.00 mL of 0.456 M $AlCl_3$, and the solution is diluted to 1.00 liter.

13. Calculate the concentration of a sulfuric acid solution if 27.89 mL of 0.9987 M sodium hydroxide is needed to neutralize completely 50.00 mL of the acid.

14. An antacid tablet contains 10.00 grams of calcium hydroxide. How many milliliters of stomach acid will one tablet neutralize? Assume stomach acid is 4.00 M HCl.

15. Calculate the concentration of each type of ion in solution after 45.67 mL of 0.9987 M $BaCl_2$ is mixed with 56.92 mL of 0.3487 M Na_2SO_4. (Assume the final volume is the sum of the initial two volumes.)

16. The amount of Vitamin C ($C_6H_8O_6$) in foods is determined by titration with iodine according to the following reaction:

$$C_6H_8O_6 + I_2 \rightarrow C_6H_6O_6 + 2\,HI$$

The Recommended Daily Allowance (RDA) for vitamin C is 60.0 mg. If 50.00 mL of fruit juice requires 27.86 mL of 0.01352 M I_2 to reach the endpoint, how many milliliters of this juice would you need to drink to meet the RDA for vitamin C?

Chapter 12: Gases

Before you get started

Many of the concepts of this chapter will be new to you. Gases behave differently than other states of matter mainly because of the distance between the molecules and the lack of intermolecular forces which might hold them together. You will learn about these reasons in Section 12.8. This chapter is longer than the last few and you should spend a proportionately longer amount of time studying it. This chapter also has more algebra than the preceding chapters. You should review how to solve a simple equation and some of the methods for relating variables by direct and indirect proportionality. Take just one section at a time and master it. Then move on to the next section.

Section 12.1: Gas Pressure

Pressure is not a new concept for you. Every time you ride a bicycle or drive a car you are depending upon the pressure in your tires to keep them round and your ride smooth. The science of weather forecasting utilizes barometric pressure to predict when fronts are moving through the atmosphere. The only new item in this section is the UNITS of pressure.

In chemistry, pressure is measured in one of two units, atmospheres or mm Hg (torr). The SI (Système International d'Unités) unit of pressure is the pascal. This is being used in many chemistry texts today. Your text is using the more familiar units. You can easily relate these units if they are presented together.

	1.00 atm (atmosphere)
	760 mm Hg (millimeters of mercury)
Equivalent Gas Pressures	760 torr
	14.7 psi (pounds per square inch)
	29.92 inches of mercury
	101,325 pascals

☐ Make sure to read the Examples and work the Practice Problems in the text as you read this section. Check your answers carefully.

☐ Work the Problems for this section at the end of the chapter.

Section 12.2: Boyle's Law

Much of the material in this chapter is more easily understood if you can relate it to things you see every day. Boyle's Law is the law that makes a tire pump work. If you push down on the handle (increase the pressure) you compress the gas in the cylinder (decrease the volume). This pushes the gas into the tire and increases the pressure in the tire. Another common use of this law is in the production of compressed gases. Deep sea divers use compressed air, which is simply the air you breathe packed under pressure in a cylinder. Welders also use compressed gases. When the volume is reduced, the pressure becomes greater. Compressed gases are much easier to transport because of the smaller container.

☐ Read Section 12.2.

Memorize the conversion factor between atmospheres and millimeters of mercury (torr).

1.00 atm = 760 mm Hg = 760 torr

☐ Make sure to read the Examples and work the Practice Problems in the text as you read this section. Check your answers carefully.

☐ Work the Problems for this section at the end of the chapter.

Section 12.3: Charles' Law

Again, common everyday experience will help you through this section. When cold temperatures arrive, you often find that your tires are low. The decreased temperature causes the volume to decrease. You can fill a balloon with air and place it in the freezer to see just how dramatic this effect is.

The Temperature Oath

I, _____, promise that, from this day forward, whenever I see a gas law problem involving temperature I will always convert that value to kelvins.

signed: _____

date: _____

You must always use kelvins as the unit of temperature in working with gas law problems. The volume of a gas is proportional only to the temperature in kelvins. Note that neither the word *degree* nor the symbol for degree (º) is used with the kelvin unit.

$$^{\circ}C + 273 = \text{kelvins}$$

☐ Read Section 12.3.

☐ Make sure to read the Examples and work the Practice Problems in the text as you read this section. Check your answers carefully.

☐ Work the Problems for this section at the end of the chapter.

Section 12.4: The Combined Gas Law

☐ Read Section 12.4.

Many students have problems with the algebra of the combined gas law. You may wish to memorize the amended equation below.

Combined Gas Law for the Algebraically-Challenged Student

$$P_1 V_1 T_2 = P_2 V_2 T_1$$

This is the result of clearing the fraction in the traditional combined gas law equation. You will be perfectly correct in using this equation. Most students find it easier because when you want to solve it for one of the variables, you need only divide both sides by what you do not want.

Example: Solve the above equation for V_2.

$$P_1 V_1 T_2 = P_2 V_2 T_1$$

To get V_2 alone, on the right side of the equation, divide both sides by $P_2 T_1$.

$$\frac{P_1 V_1 T_2}{P_2 T_1} = \frac{P_2 V_2 T_1}{P_2 T_1}$$

$$\frac{P_1 V_1 T_2}{P_2 T_1} = V_2$$

P_2 and T_1 cancel out on the right side of the equation, leaving you with the solution rather quickly. After some practice, you will be able to do this in your head with very little effort. Remember that in this equation, T_1 will always start on the same side of the equation as P_2V_2, and T_2 will always start on the same side as P_1V_1.

☐ Use this method as you read the exercises and work the Practice Problems of this section.

☐ Note the definition of **standard temperature and pressure**. This can be written several ways: 0°C is also 273 kelvins; 1 atmosphere is also 760 mm Hg, or 760 torr.

☐ Make sure to read the Examples and work the Practice Problems in the text as you read this section. Check your answers carefully.

☐ Work the Problems for this section at the end of the chapter.

Section 12.5: The Ideal Gas Law

☐ Read Section 12.5.

Whenever you see that a problem is using or asking for the number of moles or grams of gas present, you can be fairly sure that you will need to use the Ideal Gas Law.

Be very careful to always convert the temperature to kelvins. Also, remember the diatomic gases. Note that in Example 12.15, the molar mass of nitrogen is that of N_2, equal to 28.0 grams per mole.

For review, write the formulas for the seven diatomic elements you learned in Chapter 5.

_____ _____ _____ _____ _____ _____ _____

Calculate the **molar mass** for each of them: (This is twice the atomic mass because there are two atoms in each diatomic gas).

_____ _____ _____ _____ _____ _____ _____

☐ Make sure to read the Examples and work the Practice Problems in the text as you read this section. Check your answers carefully.

☐ Work the Problems for this section at the end of the chapter.

Section 12.6: Gases in Chemical Reactions

☐ Examine the chart in Figure 12.7. This relates all the conversions you have learned to this point about chemical reactions and stoichiometry. By understanding the various paths of this chart you should be able to do some very complicated chemical calculations. Take some time and look over the different paths. Make up some problems that will take you through the chart in several different ways. Use these problems to study for an exam on this material. Exchange your problems with other members of your study group and see if you can trick each other.

☐ Read Section 12.6.

☐ Read the examples and work the Practice Problems of this section. Be sure to read the narrative with these examples. There is much information in this section about putting together information you have learned in the past. Be prepared to make up some very sophisticated problems from this section.

One mole of a substance has a mass equal to the molecular mass of the substance in grams. Density is the measure of grams of substance per unit volume.

What is the density of nitrogen gas at STP?

N_2 has a molar mass of 28.0 g/mol.

$$n = \frac{m}{MM}$$

$$PV = nRT = \frac{mRT}{MM}$$

$$d = \frac{m}{V} = \frac{P(MM)}{RT}$$

$$= \frac{(1.00 \text{ atm})(28.0 \text{ g/mol})}{(0.0821 \text{ L·atm/mol·K})(273 \text{ K})}$$

$$= 1.25 \text{ g/L}$$

Density of gases is usually expressed as grams per liter.

Practice Problem 12.18 gives you a useful piece of information. The **standard molar volume** of a gas is the volume occupied by one mole of any gas **at STP**. This is **22.4 liters/mole**. The above problem becomes much simpler using this number.

What is the density of N_2 at STP?

$$\frac{28.0 \text{ g}}{\text{mole}} \times \frac{1 \text{ mole}}{22.4 \text{ L}} = 1.25 \text{ g/L}$$

Caution! You can calculate the density this way only for gases at STP.

☐ Make sure to read the Examples and work the Practice Problems in the text as you read this section. Check your answers carefully.

☐ Work the Problems for this section at the end of the chapter.

Section 12.7: Dalton's Law of Partial Pressures

A gas will have the same pressure whether it is alone in a container or combined with other gases in the same container. For this reason the total pressure in a container will be the sum of the individual pressures of each of the gases present.

☐ Read Section 12.7.

☐ Make sure to read the Examples and work the Practice Problems in the text as you read this section. Check your answers carefully.

☐ Work the Problems for this section at the end of the chapter.

Section 12.8: Kinetic Molecular Theory of Gases

You learned about kinetic energy in Section 2.6. The energy of a body in motion is related to its mass and its velocity. Kinetic molecular theory of gases allows for the understanding of how macroscopic properties of gases can correspond with the gas molecules' interactions on the submicroscopic scale.

☐ Read Section 12.8.

☐ Look carefully at the words in **bold print**. Be sure to understand how this theory can be used to explain pressure and how gases mix with one another.

☐ Make sure to read the Examples and work the Practice Problems in the text as you read this section. Check your answers carefully.

☐ Work the Problems for this section at the end of the chapter.

Chapter 12: Finishing Up

☐ Carefully read the summary section at the end of the chapter. Do you understand each paragraph? Do you know the terms used? If not, review the section indicated at the end of the paragraph.

☐ The section titled "Items for Special Attention" gives you some pointers to help you learn the material correctly and also can keep you from making some of the more common mistakes. Read these carefully.

☐ Work the "Self-Tutorial Problems." Check each answer as you complete the problem. If you have trouble with one of these, be sure to ask your instructor for help.

☐ Look over the list of "Key Terms" at the beginning of the chapter. If you do not recognize a term or are unsure of how it was used in the chapter, go back to that section and reread it.

☐ Use your lecture notes and the text to find the important topics covered in class. Make flash cards so that you can study these carefully before the exam.

☐ Work each of the problems at the end of the chapter in your text. Be sure to check your answers. If the answer is not given for that problem, work another that is similar and check the answer. If you are incorrect, check your work and review the chapter to see if you can answer the question correctly. If you cannot, get help from your instructor.

☐ Work the sample exam questions on the next few pages.

☐ Check your answers when you have completed all of them.

☐ Make up your own exam and exchange it with a fellow student or use it in your study group.

Chapter 12: Sample Exam Questions

1. In the following table, calculate the missing variables by using the combined gas law:

	V_1	P_1	T_1	V_2	P_2	T_2
Gas 1	476 mL	735 torr	298 K		790 torr	273 K
Gas 2	1.25 L	5.21 atm	240 K	5.50 L		300 K
Gas 3	6.43 L	678 torr		14.7 L	854 torr	300 K
Gas 4	729 mL		250 K	2.35 L	3.89 atm	450 K

2. How many liters of oxygen can be generated at 20°C and 0.998 atm by the decomposition of 3.89 grams of water?

3. A gas has a volume of 3.75 L at 0°C and 1.00 atm. If the volume of the gas remains constant, at what Celsius temperature would the gas have a pressure of 4.50 atm?

4. A 50.0 L balloon is filled with helium at 25°C and 753 mm Hg barometric pressure. The balloon is released and climbs to an altitude where the pressure is 359 mm Hg. What is the volume of the balloon if, during the ascent, the temperature drops to -30°C? Assume that there is no pressure exerted by the elasticity of the walls of the balloon.

5. What is the volume of 25.0 grams of nitrogen gas at 0°C and 1.50 atm?

6. An ideal gas occupies 25.8 L at 2.51 atm and 315 K. If the gas is compressed to 17.2 L and the temperature is lowered to 261 K, what is the new pressure?

7. The inside pressure of an automobile tire is 1.75 atm and the temperature is 20°C. Assuming that the volume remains constant, what will the pressure in the tire be if, after 200 miles of driving, the temperature of the tire increases to 50°C?

8. When hydrogen gas is released in the reaction of HCl and Zn, the volume of a certain sample of H_2 is 76.3 mL at 26.3°C and 757 mm Hg. What would the volume of the hydrogen be at STP?

9. A scuba tank holds 8.00 L of air at a pressure of 140 atm at 20°C. How many liters of air would this be at STP?

10. The average person inhales 500 mL of air (assume STP) 12 times every minute. Using the volume obtained in question 9, calculate how many hours a full scuba tank would provide air for a diver.

11. What volume is occupied by 6.85 grams of methane gas (CH_4) at 22°C and 1.25 atm pressure?

12. How many grams of carbon dioxide must be placed into a 50.0 L tank to develop a pressure of 1000 torr at 25°C?

13. What is the density in grams per liter of nitrogen dioxide at STP?

14. A sample of oxygen collected over water exerts a total pressure of 748 mm Hg at 22°C. The sample has a volume of 320 mL. The vapor pressure of water at 22°C is 19.8 mm Hg. How many grams of O_2 does the sample contain?

15. An organic compound has the following percentage composition: 55.8% carbon, 7.0% hydrogen, and 37.2% oxygen. If 3.26 grams of the compound occupies 1.47 L at 160°C and 0.914 atm, what is the molecular formula of the compound?

Chapter 12: Experiment at Home

The Power of Air

Air pressure surrounds you everyday. Approximately one atmosphere of pressure pushes in on you from all directions. This is about 14.7 pounds per square inch. The fluids of your body push back to compensate for the pressure. But just how much pressure is atmospheric pressure? The following experiment will demonstrate to you how much force there is around you.

Materials: safety goggles (available at most hardware stores)
clean metal can with tight fitting lid (Make sure there are no flammable solvents in the can.)
stove
hotpads or gloves
cup of water

Put on your Safety goggles.

With the lid off, put the water in the can to a depth of about one centimeter. Heat the can until the steam from the water can be seen leaving the opening steadily. Continue heating briskly for two more minutes. Using the hotpads, remove the can immediately from the heat. Quickly and carefully place the top on the can and tighten it securely. Allow the can to cool.

The steam forces the air out of the can. As the can cools, the water will condense and become a liquid again. This will decrease the pressure in the can to a point much below that outside of the can. The can will collapse under the outside pressure.

Chapter 13: Atomic and Molecular Properties

Before you get started

☐ Review Chapter 3. Remember the basic principles about the atom and the subatomic particles.

Particle	Location	Charge	Mass
proton	in nucleus	+1	1 amu
neutron	in nucleus	0	1 amu
electron	outside nucleus	-1	0 amu

☐ You may want to have a periodic table handy.

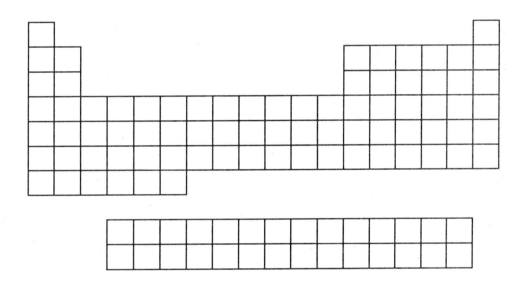

Section 13.1: Atomic and Ionic Sizes

Circle the word that correctly completes each sentence..

1. Particles with (the same) (different) charges attract each other.

2. Like charged particles (repel) (attract) each other.

3. The closer the two charged bodies, the (greater) (less) the force of attraction or repulsion.

4. Electrons are (heavy)(light) and lie (inside)(outside) of the nucleus, while protons are (heavy)(light) and lie (inside)(outside) of the nucleus.

5. Ions formed by the gain of one or more electrons are (larger)(smaller) than the original atom.

6. Ions formed by the loss of one or more electrons are (larger)(smaller) than the original atom.

☐ Make sure to read the Examples and work the Practice Problems in the text as you read this section. Check your answers carefully.

☐ Work the Problems for this section at the end of the chapter.

Section 13.2: Ionization Energy and Electron Affinity

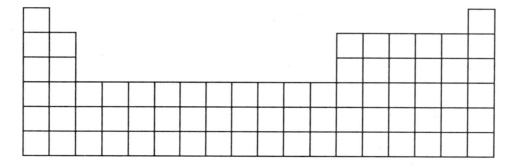

The general trend is for the atomic radius to *decrease* across a row and to *increase* down a column of the periodic table. The general trend for the first ionization energy of an atom is to *increase* across a row and *decrease* down a column of the periodic table. Therefore, the elements with the *smallest* atomic radii are found in the upper right hand corner of the table The elements with the *highest* first ionization energy are also found in the upper right hand corner of the periodic table. Conversely, the elements with the *largest* atomic radii are found in the lower left hand corner of the table. The elements with the *lowest* first ionization energy are also found in the lower left hand corner of the periodic table. (Note: Hydrogen has the smallest atomic radius.)

Electron affinity is the amount of energy given off when an electron is added to an atom to form an ion in the gaseous state. The smaller the atom, the stronger the attraction for the electron (first ionization energy) and the more energy is given off when the electron is added to the atom. If you remember that the smallest atoms are in the upper right hand part of the periodic table, then you will remember that this is where the atoms also have the highest electron affinity.

☐ Place arrows beside the periodic table above to show increasing atomic radius, increasing first ionization energy, and increasing electron affinity

☐ Make sure to read the Examples and work the Practice Problems in the text as you read this section. Check your answers carefully.

☐ Work the Problems for this section at the end of the chapter.

Section 13.3: Electronegativity and Bond Polarity

Electronegativity is a measure of the ability of an atom to hold electrons in a bond. If the two atoms forming the bond have the same electronegativity as in the case of a diatomic element, the difference is zero and the bond formed is 100% covalent. The electrons are shared equally between the two atoms. If there is any difference in electronegativity, the electrons will be held closer to the atom with the higher electronegativity. The greater the difference in electronegativity, the more ionic character the bond will possess.

☐ Make sure to read the Examples and work the Practice Problems in the text as you read this section. Check your answers carefully.

☐ Work the Problems for this section at the end of the chapter.

Section 13.4: Molecular Shape

Molecular geometry is determined by the shape of the *atoms* in a molecule. The arrangement of bonding and nonbonding *electrons* determines where the atoms will be located in the molecule.

Since the electrons surrounding a central atom in a molecule are all negatively charged, it is expected that they will try to occupy positions that are as far apart as possible from each other and still remain attached to the central atom. The geometry providing the furthest separation of the electron clouds is important. Blow up four balloons that are about the same size and shape. Tie two of the balloons together and observe the geometry. It should appear linear, or a straight line. When a central atom has *two* electron clouds around it the resulting shape will be ***linear***.

When the third balloon is tied to the first two, the shape becomes triangular. Note that the three balloons will be coplanar, that is, they will all be in the same plane. Thus, when a central atom has *three* electron clouds around it the resulting shape will be ***triangular***. Note that this geometry is coplanar; all of the atoms are in a flat configuration.

When the fourth balloon is tied to the three, the shape becomes tetrahedral (a four cornered pyramid). When a central atom has *four* electron clouds around it the resulting shape will be ***tetrahedral***.

It doesn't matter whether the clouds are bonding or nonbonding pairs of electrons, the geometry is based upon the number of electron clouds.

The molecular geometry is then determined by the arrangement of the atoms in the molecule without regard to the non-bonding pairs of electrons. If there are three electron clouds the shape of the clouds is triangular. But if one of those clouds is a nonbonding pair of electrons, the molecular geometry is determined to be bent or nonlinear because that is what is seen in the shape of the molecule.

Similarly, if there are four electron clouds the shape of the clouds is tetrahedral. But if one of those clouds is a nonbonding pair of electrons, the molecular geometry is trigonal pyramidal. (Do not confuse this with triangular. The central atom in the trigonal pyramidal is not in the same plane as the three legs.) If two clouds are nonbonding pairs of electrons, the molecular geometry is bent or nonlinear.

Number of Electron Clouds	Electron Pair Geometry	Number of Bonding Electron Pairs	Number of Non-Bonding Electron Pairs	Molecular Geometry
2	2	2	0	linear
		1	1	linear
3	3	3	0	triangular
		2	1	bent
		1	2	linear
4	4	4	0	tetrahedral
		3	1	trigonal pyramidal
		2	2	bent
		1	3	linear

☐ Make sure to read the Examples and work the Practice Problems in the text as you read this section. Check your answers carefully.

☐ Work the Problems for this section at the end of the chapter.

Section 13.5: Polar and Nonpolar Molecules

An easy way to visualize the concept of molecular polarity is to imagine a balanced teeter-totter. It doesn't matter how much weight is on each end as long as the weights are approximately the same. If there is much more weight on one end than the other, no attempt at balance will be successful. If we consider the bond polarity as the "weight," we can see that even a molecule with highly polar bonds can be nonpolar if all the bonds "balance" each other.

☐ Make sure to read the Examples and work the Practice Problems in the text as you read this section. Check your answers carefully.

☐ Work the Problems for this section at the end of the chapter.

Section 13.6: Intermolecular Forces

In addition to the chemical bonds which hold atoms together in molecules or ionic compounds, there are three important intermolecular forces that cause molecules to be attracted to each other. None of these is as strong as a covalent bond, but they do play important roles in the chemistry of liquids and molecular solids.

The weakest of these is **van der Waals forces**. This is sometimes called an *induced dipole* because the electrons around one molecule are affected by the electrons of another molecule and form a very weak dipole.

Dipole moments are actual charge separations in molecules. You learned how to predict molecular polarity in Section 13.5. Polar molecules attract each other much as magnets attract.

The strongest of the intermolecular forces is **hydrogen bonding.**

☐ Make sure to read the Examples and work the Practice Problems in the text as you read this section. Check your answers carefully.

☐ Work the Problems for this section at the end of the chapter.

Chapter 13: Finishing Up

☐ Carefully read the summary section at the end of the chapter. Do you understand each paragraph? Do you know the terms used? If not, review the section indicated at the end of the paragraph.

☐ The section titled "Items for Special Attention" gives you some pointers to help you learn the material correctly and also can keep you from making some of the more common mistakes. Read these carefully.

☐ Work the "Self-Tutorial Problems." Check each answer as you complete the problem. If you have trouble with one of these, be sure to ask your instructor for help.

☐ Look over the list of "Key Terms" at the beginning of the chapter. If you do not recognize a term or are unsure of how it was used in the chapter, go back to that section and reread it.

☐ Use your lecture notes and the text to find the important topics covered in class. Make flash cards so that you can study these carefully before the exam.

☐ Work each of the problems at the end of the chapter in your text. Be sure to check your answers. If the answer is not given for that problem, work another that is similar and check the answer. If you are incorrect, check your work and review the chapter to see if you can answer the question correctly. If you cannot, get help from your instructor.

☐ Work the sample exam questions on the next few pages.

☐ Check your answers when you have completed all of them.

☐ Make up your own exam and exchange it with a fellow student or use it in your study group.

Chapter 13: Sample Exam Questions

1. Which of the following elements has the *largest* atomic radius?

    S   Al   Mg   K   Si

2. Which of the following elements has the *smallest* atomic radius?

    Sr   Ca   Mg   Ba   Be

3. Which of the following elements has the *highest* first ionization energy?

    S   Al   Mg   K   Si

4. Which of the following elements has the *lowest* first ionization energy?

    Sr   Ca   Mg   Ba   Be

5. For each of the following pairs, circle the one with the LARGER radius:

    Na Na^+

    Cl Cl^-

    Mg^{2+} Mg

    O^{2-} O

6. Which of the following elements has the *highest* electronegativity?

    S   Al   Mg   P   Si

7. Which of the following elements has the *lowest* electronegativity?

    Sr   Ca   Mg   Ba   Be

8. Using Figure 5.1, determine the difference in electronegativity for each of the following bonds:

        Li   and   F   _____

        K   and   O   _____

        C   and   H   _____

        C   and   N   _____

        Si   and   O   _____

Arrange these bonds in increasing order of polarity.

9. Predict the *molecular* geometry for each of the following: (You will need to draw the Lewis structure for each)

        SO_2 _____

        SO_3 _____

        CO_2 _____

        H_2S _____

        NH_3 _____

10. For each molecule in Problem 9, predict the overall molecular polarity:

        SO_2 _____

        SO_3 _____

        CO_2 _____

        H_2S _____

        NH_3 _____

11. NO and CO have approximately the same molecular weight. Which compound has the greater intermolecular attractions. (HINT: Remember what you learned about electronegativity.)

Chapter 14: Solids and Liquids, Energies of Physical and Chemical Changes

Matter is held together by several different types of forces. In this chapter you will learn how these forces can affect such things as melting point and boiling point. Much of what you will learn in this chapter is easily related to familiar substances and processes. Think for a moment about water. This very common substance is composed of molecules with a molar mass of only 18 g/mol. Carbon dioxide has a molar mass of 44 g/mol. Carbon dioxide is a gas and water is a liquid or solid at STP. It is because of the strong intermolecular forces in water that the water is a liquid. Water also provides a good example of state change. You are familiar with ice, water, and steam. You know that as you add heat to ice, it melts, and if you add energy to water on the stove it boils to form steam.

☐ Look at Table 14.1. As you read about the properties of the particles in each of the states of matter, think about ice, water, and steam.

Section 14.1: Nature of the Solid and Liquid States

Individual samples of gold, silver, diamond, and many other solids have been around for many centuries. Solids represent the most organized form of matter. Atoms are held rigidly into a matrix which is predetermined by the nature of the atom. Many solids can be identified by the shape of the crystal they form. Quartz is easily recognized by the angles of its crystal faces which reflect the silicon-oxygen bonds within. Sodium chloride forms a crystal which is a cube. In fact, the salt in your salt shaker will be the same cubic structure. Diamond has a very distinctive shape. Water crystallizes into hexagonal shapes. Look closely at snowflakes to see the six-sided shapes.

☐ Read Section 14.1.

Table 14.2 describes the types of crystalline solids. The stronger the intermolecular forces between the molecules of a substance are, the higher the boiling point of the substance will be. An ionic solid is held together by electrostatic charges in a lattice structure. A molecular substance will have van der Waal forces, dipole-dipole interactions, and/or hydrogen bonding holding the molecules together in the liquid phase. Ionic substances have higher melting points than molecular substances.

☐ Make sure to read the Examples and work the Practice Problems in the text as you read this section. Check your answers carefully.

☐ Work the Problems for this section at the end of the chapter.

Section 14.2: Changes of Phase

☐ You should know the terminology associated with phase changes. Fill in the following blanks using these words:

When a solid is converted to a liquid the process is called _____.

When a solid is converted to a gas the process is called _____.

When a liquid is converted to a solid the process is called _____.

When a liquid is converted to a gas the process is called _____.

When a gas is converted to a liquid the process is called _____.

When a gas is converted to a solid the process is called _____.

☐ Note carefully the words in **bold print**. Physical equilibrium is attained when there is no more change. The rate of condensation is equal to the rate of evaporation. No change of volume of the liquid can be detected. The pressure above the liquid is also constant at this point and is called the *vapor pressure*. You can review this concept in Chapter 12. Be sure you understand the difference between boiling point and normal boiling point.

☐ Make sure to read the Examples and work the Practice Problems in the text as you read this section. Check your answers carefully.

☐ Work the Problems for this section at the end of the chapter.

Section 14.3: Energy Changes

The energy associated with any change is calculated by taking the initial energy and subtracting it from the final energy. This may sometimes result in a negative number but, as you will see later, the order is important,. The Greek letter delta, Δ, is used in chemistry to mean the **final** minus **initial**.

$$\Delta = \text{final} - \text{initial}$$

☐ Memorize the formula:

$$\text{Heat} = mc\Delta t$$

The heat required to change the temperature of m grams of a substance Δt degrees Celsius is found by multiplying the mass of the substance times the specific heat of the substance times the change in temperature.

Specific heats of some substances are listed in Table 14.4.

Many textbooks use the specific heat of water in calories instead of joules. Since the definition of a calorie is the amount of energy necessary to raise the temperature of one gram of water one degree Celsius, then the specific heat of water is 1.00 cal/g·°C. Be sure to carefully check the units when you perform calculations.

NOTE: Use the factor label method when working these problems. The units of specific heat will tell you that in order to get an answer in joules, you must multiply it by mass **and** temperature. To calculate the energy necessary for a given phase change you only need to know the mass of the substance since there is no temperature change involved.

☐ Table 14.4 shows the values of specific heats of various substances. These are used to calculate the amount of energy required to change the **temperature** of a substance. Table 14.5 shows the values of heats of phase change. The energy required to change one gram of a substance from the solid to the liquid phase is called the Heat of Fusion, ΔH_{fus}. The energy required to change one gram of a substance from the liquid to the gas phase is called the Heat of Vaporization, ΔH_{vap}.

When you read a heating curve such as the one in Figure 14.6, the following steps should be observed:

Warming of solid

To calculate the energy needed to warm a substance, you need to know the mass of the substance, how many degrees the temperature must be increased to reach the phase change temperature, and the specific heat of the solid (Table 14.4).

$$\text{Heat} = mc\Delta t$$

Melting

The amount of energy necessary to melt the substance is calculated by knowing the heat of fusion of the substance (Table 14.5) and the mass of the substance.

$$\text{Heat} = m\, \Delta H_{fus}$$

Warming of liquid

This is calculated with the first equation. Again you will need to know the mass of the substance, how many degrees the temperature is to be increased, and the specific heat of the liquid (Table 14.4).

$$\text{Heat} = mc\Delta t$$

Boiling

The energy required to boil a liquid is calculated by knowing the heat of vaporization of the liquid (Table 14.5) and the mass of the liquid.

$$\text{Heat} = m\, \Delta H_{vap}$$

Warming of vapor

The energy required to warm the gas is calculated with the first equation. Again, you will need to know the mass of the gas, how many degrees the temperature will be increased, and the specific heat of the gas (Table 14.4).

$$\text{Heat} = mc\Delta t$$

After all of these values are calculated, they should each be in units of joules. They can then be added together to obtain the total energy required. Sometimes you may be asked to find the energy in calories. The same steps are used, but you need to use the specific heats and heats of fusion and vaporization in calories instead of joules. Do not mix up these units.

This process can also be reversed to calculate the total energy that needs to be removed in order to change a substance from gas to solid.

☐ Make sure to read the Examples and work the Practice Problems in the text as you read this section. Check your answers carefully.

☐ Work the Problems for this section at the end of the chapter.

Section 14.4: Enthalpy Changes in Chemical Reactions

When energy, heat or work is added to a system, the energy change is positive. When energy, heat or work is removed from a system, the energy change is negative. The sign of the energy is very important in doing the calculations of this chapter; keep the sign with the values you use.

Also remember the "delta" sign has the following meaning:

$$\Delta = \text{final} - \text{initial}$$

$$\Delta = \text{products} - \text{reactants}$$

There are many kinds of enthalpy changes associated with various processes. The enthalpy change of a substance being formed from its elements, all in their standard or most stable states, is called the enthalpy of formation, ΔH_f.

What to remember when performing enthalpy of formation calculations:

1. Balance the equation.
2. Uncombined elements in the standard state have an enthalpy of formation of zero.
3. Some elements are diatomic.
4. Carefully record the sign of each value you look up or calculate.
5. Remember to use the stoichiometric coefficients when doing calculations.

Hess's Law is used to determine the ΔH for a reaction indirectly. Sometimes the enthalpy of formation for a reactant or product cannot be found in a table. By adding together ΔH values for several reactions for which the enthalpy data is known, the missing values can be calculated.

☐ Make sure to read the Examples and work the Practice Problems in the text as you read this section. Check your answers carefully.

☐ Work the Problems for this section at the end of the chapter.

Chapter 14: Finishing Up

☐ Carefully read the summary section at the end of the chapter. Do you understand each paragraph? Do you know the terms used? If not, review the section indicated at the end of the paragraph.

☐ The section titled "Items for Special Attention" gives you some pointers to help you learn the material correctly and also can keep you from making some of the more common mistakes. Read these carefully.

☐ Work the "Self-Tutorial Problems." Check each answer as you complete the problem. If you have trouble with one of these, be sure to ask your instructor for help.

☐ Look over the list of "Key Terms" at the beginning of the chapter. If you do not recognize a term or are unsure of how it was used in the chapter, go back to that section and reread it.

☐ Use your lecture notes and the text to find the important topics covered in class. Make flash cards so that you can study these carefully before the exam.

☐ Work each of the problems at the end of the chapter in your text. Be sure to check your answers. If the answer is not given for that problem, work another that is similar and check the answer. If you are incorrect, check your work and review the chapter to see if you can answer the question correctly. If you cannot, get help from your instructor.

☐ Work the sample exam questions on the next few pages.

☐ Check your answers when you have completed all of them.

☐ Make up your own exam and exchange it with a fellow student or use it in your study group.

Chapter 14: Sample Exam Questions

1. The heat of vaporization for ethanol is 0.880 kJ/g. If 96.4 kJ is released by the condensation of a sample of ethanol, what is the mass of the sample?

2. Given the following information:

$$\text{specific heat of ice} = 2.09 \text{ J/g} \cdot {}^{\circ}\text{C}$$
$$\text{specific heat of water} = 4.184 \text{ J/g} \cdot {}^{\circ}\text{C}$$
$$\text{heat of fusion of ice} = 335 \text{ J/g}$$

calculate the amount of energy required to raise the temperature of 13.98 grams of ice at -25.0°C to water at +25.0°C.

3. The heat of vaporization for ethanol is 0.880 kJ/g. If 45.9 kJ is required to evaporate a sample of ethanol, what is the mass of the sample?

4. A substance gives off 35.7 kJ of heat when 11.7 grams of the substance condenses. What is the heat of vaporization of the substance?

5. Acetone, C_3H_6O, is a major component in fingernail polish remover. It makes the skin feel cold when it evaporates. How much heat is absorbed when 2.98 grams of acetone evaporates from the skin? The heat of vaporization is 32.0 kJ/mol.

6. Find the amount of heat given off when 58.2 kg of iron is drawn from a blast furnace at 1645°C and poured into a mold, cooled, solidified, and then cooled further to room temperature (26°C). Iron melts at 1535°C. Liquid iron has an average specific heat of 0.452 J/g·°C. The heat of fusion of iron is 267 J/g. The average specific heat of solid iron is 0.444 J/g·°C.

7. Fill in the proper names for the following state changes:

    (a) liquid to solid _____

    (b) solid to gas _____

    (c) liquid to gas _____

8. For most substances, heats of vaporization are much larger than heats of fusion. Water, for example, has a heat of vaporization of 2260 J/g and a heat of fusion of 335 J/g. Explain.

Chapter 15: Solutions

Polarity of molecules is dependent upon the symmetry of the molecules. If there is balance, the molecule is most likely non-polar. If the molecule is asymmetric because of lone pairs of electrons, then the asymmetry will result in polarity. Look back to Chapter 5 where you learned how to draw electron dot structures for molecules. You will be able to predict the polarity of a molecule by looking at the types of bonds formed and how symmetrical they are.

Some examples of ionic compounds are:

$$NaCl \quad AgNO_3 \quad NaOH \quad MgCl_2 \quad ZnCl_2$$

NOTE: There are ionic substances which are not soluble in water. You learned about these in Chapter 9. Review these at this time.

Some examples of covalent compounds which are asymmetric and therefore polar are:

$$CH_3CH_2OH \quad\quad CH_3COOH \quad\quad CH_3CH_2NH_2$$

Some examples of covalent compounds which are symmetric and therefore nonpolar are:

$$C_6H_6 \quad\quad CH_3CH_2CH_2CH_3 \quad\quad CO_2$$

☐ Draw the electron dot structure for water, H_2O. Can you see the two lone pairs of electrons? Draw a circle around the entire molecule. On one side of the circle you will see the oxygen atom with the lone pairs of electrons. On the other side of the circle there are *two* hydrogen atoms. The hydrogen end of the water molecule is the positive pole of the dipole; the oxygen end is the negative pole.

Section 15.1: The Solution Process

☐ Read Section 15.1.

Remember that "like dissolves like" refers to the polarity of the solute and solvent molecules.

☐ Make sure to read the Examples and work the Practice Problems in the text as you read this section. Check your answers carefully.

☐ Work the Problems for this section at the end of the chapter.

Section 15.2: Saturated, Unsaturated, and Supersaturated Solutions

☐ Read Section 15.2.

☐ Define the following words that appear in this section.

        solubility: _____

        saturated solution: _____

        unsaturated solution: _____

        supersaturated solution: _____

Solubility is related to temperature. Look at Figure 15.2. Most of the substances become more soluble as the temperature increases.

☐ Make sure to read the Examples and work the Practice Problems in the text as you read this section. Check your answers carefully.

☐ Work the Problems for this section at the end of the chapter.

Section 15.3: Molality

☐ Read Section 15.3.

Be very careful of the spelling and the definition of molality. They are very close to molarity.

$$m = \frac{\text{moles of solute}}{\text{kilograms of solvent}}$$

NOTE: The denominator of the definition for molarity is *liters of solution*, for molality is *kilograms of solvent*.

☐ Make sure to read the Examples and work the Practice Problems in the text as you read this section. Check your answers carefully.

☐ Work the Problems for this section at the end of the chapter.

Section 15.4: Mole Fraction

You will be using mole fraction to perform calculations in the next section.

☐ Read Section 15.4.

☐ Make sure to read the Examples and work the Practice Problems in the text as you read this section. Check your answers carefully.

☐ Work the Problems for this section at the end of the chapter.

Section 15.5: Colligative Properties

☐ Find the definition of *colligative* and write it here:

_____

☐ Write the names of the four colligative properties listed in your text:

_____

_____

_____

_____

☐ Read the section under *Vapor-Pressure Lowering*.

The vapor pressure of a pure solvent is lowered by the addition of a solute. The normal boiling point of a substance is the temperature at which the vapor pressure is equal to atmospheric pressure. With added nonvolatile solute, the vapor pressure of water at 100°C is less than 760 mm Hg, and the water does not boil.

The equation used to calculate this is:

$$P_{component} = X_{component} \cdot P^o_{pure\ substance}$$

In order to perform this calculation you will need to be given the vapor pressure of the pure solvent, P^o. X is the mole fraction of the solvent.

☐ Read the section under *Freezing-Point Depression*.

The freezing point of a pure solvent is lowered by the addition of a solute.

The equation used to calculate this is:

$$\Delta t_f = k_f m$$

Δt_f is the difference between the normal freezing point for the pure substance and the freezing point after the solute has been added. The value of k_f is a constant for each solvent. The concentration of the solute must be expressed in the units of **molality**.

☐ Read the section under *Boiling-Point Elevation*.

The boiling point of a pure solvent is elevated by the addition of a solute.

The equation used to calculate this is:

$$\Delta t_b = k_b m$$

Δt_b is the difference between the boiling point of the pure substance and the boiling point after a nonvolatile solute has been added. The value of k_b is a constant for each solvent. The concentration of the solute must be expressed in the units of **molality**.

Antifreeze/coolant can be marketed as such because the same colligative properties that keep the radiator from freezing in the winter keep the radiator from boiling over in the summer.

☐ Read the section under *Osmotic Pressure*.

Osmotic pressure is the pressure required to keep solvent particles from crossing a semipermeable membrane. Only the **solvent** can move across the membrane. Therefore, the level of the side that was initially more concentrated will rise. A pressure on that side, enough to prevent the rise, is the osmotic pressure.

The equation used to calculate this is:

$$\pi V = nRT$$

π is the osmotic pressure in atmospheres, V is the volume of the solution in liters, n is the number of moles of solute, R is the gas constant (0.0821 L·atm/mol·K), and T is the temperature in kelvins.

You can also use molarity in this equation. By dividing the number of moles, n, by the volume of the solution in liters, V, you will obtain the **molarity** of the solution, M. The equation then becomes:

$$\pi = MRT$$

Be sure to always have your values in the correct units. You can check for this by always writing out the units and then cancelling them when you perform the calculation. Your answer should be in the correct units.

☐ Make sure to read the Examples and work the Practice Problems in the text as you read this section. Check your answers carefully.

☐ Work the Problems for this section at the end of the chapter.

Chapter 15: Finishing Up

☐ Carefully read the summary section at the end of the chapter. Do you understand each paragraph? Do you know the terms used? If not, review the section indicated at the end of the paragraph.

☐ The section titled "Items for Special Attention" gives you some pointers to help you learn the material correctly and also can keep you from making some of the more common mistakes. Read these carefully.

☐ Work the "Self-Tutorial Problems." Check each answer as you complete the problem. If you have trouble with one of these, be sure to ask your instructor for help.

☐ Look over the list of "Key Terms" at the beginning of the chapter. If you do not recognize a term or are unsure of how it was used in the chapter, go back to that section and reread it.

☐ Use your lecture notes and the text to find the important topics covered in class. **Make flash cards** so that you can study these carefully before the exam.

☐ Work each of the problems at the end of the chapter in your text. Be sure to check your answers. If the answer is not given for that problem, work another that is similar and check the answer. If you are incorrect, check your work and review the chapter to see if you can answer the question correctly. If you cannot, get help from your instructor.

☐ Work the sample exam questions on the next few pages.

☐ Check your answers when you have completed all of them.

☐ Make up your own exam and exchange it with a fellow student or use it in your study group.

Chapter 15: Sample Exam Questions

1. Water is a polar solvent; gasoline is a nonpolar solvent. For each compound, determine the better solvent: (Write W for water or G for gasoline in the blank.)

   _____ (a) NaCl

   _____ (b) C_6H_{14}

   _____ (c) Na_2SO_4

2. What is the molarity of a solution prepared by diluting a solution containing 34.8 grams of potassium chloride to 375 mL? (Be careful of the concentration units!)

3. A 5.00 gram sample of a nonionic, nonvolatile solute is dissolved in 50.0 g of water. The solution freezes at –2.54°C. What is the molar mass of the solute?
(k_f for water is 1.86°C/m)

4. Calculate the molality of a solution of 24.56 grams of NaCl in 250.0 grams of water.

5. Calculate the osmotic pressure of a 0.45 M solution of glucose in water at 25°C.

6. Explain why $CH_3CH_2NH_2$ is soluble in water and $CH_3CH_2CH_3$ is not.

7. Calculate the molality of a solution containing 10.00 moles of solute in 300.0 kilograms of solution.

8. How many moles of solute should be dissolved in 4.76 kg of solvent to make a solution that has a concentration of 2.35 m?

9. What is the mole fraction of each compound in a solution containing 2.14 mol H_2O and 0.987 mol CH_3CH_2OH?

10. What is the molality of the CH_3CH_2OH in Problem 9?

11. Calculate the molality of a 1.822 M NaCl solution that has a density of 1.078 g/mL.

12. Nicotine has the empirical formula C_5H_7N. A solution of 0.50 grams of nicotine in 12 grams of water boils at 100.13°C at 760 mm Hg. What is the molecular formula of nicotine?
(k_b = 0.512 °C/m)

13. When 3.26 x 10^{-3} mol of naphthalene is dissolved in 9.83 grams of benzene the freezing point of the benzene is lowered to 3.8°C (The freezing point of pure benzene is 5.5°C). What is the freezing point depression constant for benzene?

Chapter 15: Experiment at Home

Inside-Out Bubbles

The cell walls of most biological organisms consist of very thin hydrophobic (water-hating) films that are held rigid by proteins and lipids. The following experiment describes the creation of one such peculiar film.

Materials: Glass of water at room temperature
Large diameter plastic soda straw
1/2 teaspoon of clear liquid dish detergent

Optional: piece of paper
food coloring

Method: Mix the detergent into the full glass of water, trying not to stir in too much air. Let the liquid set until the surface is clear of bubbles or sweep the surface with a piece of paper to remove the stray air bubbles. Insert the soda straw into the liquid to a depth of 5/8" and place your finger over the top end of the straw to hold the liquid in the straw as you raise it out of the solution. Raise the straw to about 1/4" above the surface and release your finger. The liquid that was in the straw will, with some practice, penetrate the surface of the liquid in the glass resulting in an *inside-out bubble*, a bubble of liquid surrounded by a very thin film of air and completely immersed in the liquid. The fact that the bubble contains liquid gives it a higher density than a bubble containing air. You will see it sink to the bottom of the glass and then slowly rise to the surface where it will come to rest at a point just below the surface. The confined droplet is known as a *boule* and is generally unstable, in this case bursting after several seconds, leaving behind only a very small air bubble.

Theory: The molecules of many gases including oxygen, nitrogen and carbon dioxide, the major components of air, are hydrophobic and generally nonpolar. Hydrophobic molecules, such as the lipids that make up many biological membranes, do not mix well with water. Detergents, by contrast, consist of larger molecules with two distinct molecular regions, a hydrophilic and often ionic region that dissolves readily in water and a hydrophobic region that shows an affinity for and dissolves in other hydrophobic substances. Because of this bipolarity, detergents are soluble in both polar and nonpolar substances.

In this case the detergent serves two roles. First, it reduces the surface tension of the liquid, in essence reducing the elasticity of the surface so that the surface does not rupture or distort due to excessive internal forces. Second, it provides a distinct interface layer, somewhat like those found in cell walls where the molecules in the membranes are maintained in a stable and distinctly oriented manner.

Other projects: Place a small amount of food coloring in a second glass of water. Fill the straw with the colored water and drop the water into the uncolored water containing the detergent. This will enable you to see the inside of the bubble.

Carefully measure the diameter of one of the boules by holding a ruler to the side of the glass. Detonate the boule while underwater. From the estimated diameter of the small air bubble that results, calculate the thickness of the boule air envelope.

$$\text{Volume of sphere} = (4/3)\pi r^3$$

Chapter 16: Oxidation Numbers

Section 16.1: Assigning Oxidation Numbers

You will need to be able to assign oxidation numbers so that you can balance oxidation-reduction equations later in this chapter.

☐ Read Section 16.1.

☐ Write out the eight rules found for assigning oxidation numbers found in this section. This will help you remember them and also give you a copy to use as you work the problems.

☐ Make sure to read the Examples and work the Practice Problems in the text as you read this section. Check your answers carefully.

☐ Work the Problems for this section at the end of the chapter.

Section 16.2: Using Oxidation Numbers in Naming Compounds

☐ Read Section 16.2.

Using the Stock System for naming the nonmetal-nonmetal compounds will make the job of identifying changes in oxidation number much easier. This is what you will be doing in Section 16.4.

☐ Make sure to read the Examples and work the Practice Problems in the text as you read this section. Check your answers carefully.

☐ Work the Problems for this section at the end of the chapter.

Section 16.3: Periodic Variation of Oxidation Numbers

☐ Read Section 16.3.

On the periodic chart below, write in the names and tell the oxidation numbers of groups of elements.

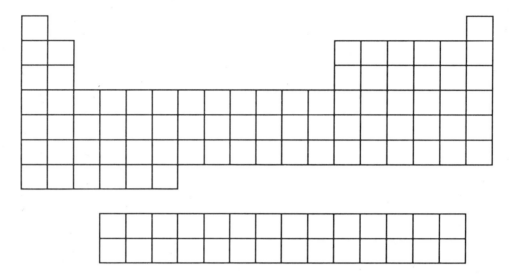

Look at Figure 16.2 as you read the following table:

Element	Group Number	Maximum Oxidation Number	Minimum Oxidation Number	Common Compounds
Na	IA	+1		NaCl
K	IA	+1		KCl
Mg	IIA	+2		$MgCl_2$
Ca	IIA	+2		$CaCl_2$
Cl	VIIA	+7	-1	NaCl (-1) NaClO (+1) $NaClO_2$ (+3) $NaClO_3$ (+5) $NaClO_4$ (+7)
S	VIA	+6	-2	Na_2S (-2) SO_2 (+4) SO_3 (+6)
N	VA	+5	-3	NH_3 (-3) N_2O_5 (+5)

☐ Make sure to read the Examples and work the Practice Problems in the text as you read this section. Check your answers carefully.

☐ Work the Problems for this section at the end of the chapter.

Section 16.4: Balancing Oxidation-Reduction Equations

When electrons are transferred from one element to another, the process involves one element becoming oxidized and the other becoming reduced. These processes always occur together. If one substance is oxidized, another substance is reduced.

☐ Read Section 16.4.

☐ Define the following as they appear in this section:

oxidation: _____

reduction: _____

oxidizing agent: _____

reducing agent: _____

There are several ways to remember the meaning of oxidation and reduction. One way is to see that whenever something is oxidized, the oxidation number becomes larger. When something is reduced, the oxidation number becomes smaller. It will be helpful also if you can determine whether electrons are gained or lost. For this reason many students use the following:

Loss of Electrons is Oxidation. Gain of Electrons is Reduction.

LEO GER

Leo, the lion says, "ger."

In a redox reaction, one substance becomes reduced. It is called the oxidizing agent because, by gaining electrons and becoming reduced, it caused something else to become oxidized, or lose electrons. The reducing agent causes something else to be reduced (gain electrons) and itself becomes oxidized (loses electrons).

Redox reactions may be broken into two half reactions: the oxidizing half reaction where electrons are on the product side (right side), and the reducing half reaction where electrons are on the reactant side (left side).

Remember that the number of electrons lost in oxidation must equal the number of electrons gained in reduction.

These reactions can be balanced in either acid or base solution. In acid there are H^+ ions present. When balancing the equation, add the H^+ to the side that has an excess of oxygen and H_2O to the side with less oxygen.

To balance a redox reaction in basic solution, simply add OH^- to both sides of the equation. Where there is H^+, there will now be H_2O. The H_2O on both sides can be cancelled.

For example,

$$MnO_4^- + 4\ H^+ + 3\ e^- \rightarrow MnO_2 + 2\ H_2O$$

Add 4 OH^- to both sides (the same as the number of H^+):

$$4\ OH^- + 6\ MnO_4^- + 4\ H^+ + 3\ e^- \rightarrow MnO_2 + 2\ H_2O + 4\ OH^-$$

Adding the H^+ and the OH^- gives 4 H_2O:

$$4\ H_2O + MnO_4^- + 3\ e^- \rightarrow MnO_2 + 2\ H_2O + 4\ OH^-$$

Cancelling 2 H_2O on both sides leaves the final answer in base solution:

$$2\ H_2O + MnO_4^- + 3\ e^- \rightarrow MnO_2 + 4\ OH^-$$

☐ Make sure to read the Examples and work the Practice Problems in the text as you read this section. Check your answers carefully.

☐ Work the Problems for this section at the end of the chapter.

Section 16.5: Equivalents and Normality

☐ Read Section 16.5.

☐ Write the two definitions of an equivalent:

_____

_____

Normality is defined as equivalents of solute per liter of solution.

The equivalent mass is the number of grams per equivalent. If you know what the substance is, you can use the atomic or molecular mass and divide that by the number of equivalents in each mole.

☐ Make sure to read the Examples and work the Practice Problems in the text as you read this section. Check your answers carefully.

☐ Work the Problems for this section at the end of the chapter.

Chapter 16: Finishing Up

☐ Carefully read the summary section at the end of the chapter. Do you understand each paragraph? Do you know the terms used? If not, review the section indicated at the end of the paragraph.

☐ The section titled "Items for Special Attention" gives you some pointers to help you learn the material correctly and also can keep you from making some of the more common mistakes. Read these carefully.

☐ Work the "Self-Tutorial Problems." Check each answer as you complete the problem. If you have trouble with one of these, be sure to ask your instructor for help.

☐ Look over the list of "Key Terms" at the beginning of the chapter. If you do not recognize a term or are unsure of how it was used in the chapter, go back to that section and reread it.

☐ Use your lecture notes and the text to find the important topics covered in class. Make flash cards so that you can study these carefully before the exam.

☐ Work each of the problems at the end of the chapter in your text. Be sure to check your answers. If the answer is not given for that problem, work another that is similar and check the answer. If you are incorrect, check your work and review the chapter to see if you can answer the question correctly. If you cannot, get help from your instructor.

☐ Work the sample exam questions on the next few pages.

☐ Check your answers when you have completed all of them.

☐ Make up your own exam and exchange it with a fellow student or use it in your study group.

Chapter 16: Sample Exam Questions

1. Balance the following redox reaction in acid solution:

$$MnO_4^- + H_2S \rightarrow MnO_2 + S$$

2. An aqueous solution of 10.0%(w/v) KCl is used for treating potassium deficiency. How many meq of K^+ are there in a 25.0 mL dose? [10.0%(w/v) means that there are 10.0 grams of solute in 100 mL of solution.]

3. Calculate the number of equivalents of solid acid that can be neutralized by 24.89 mL of 0.08763 N NaOH.

4. Complete and balance an equation for the following redox reaction in acid solution:

$$S_2O_3^{2-} + Cl_2 \rightarrow SO_4^{2-} + Cl^-$$

5. What is the concentration of a solution of 0.075 M H_2SO_4 in eq/L?

6. Calculate the normality of an acid if 20.00 mL of the acid is neutralized by 14.59 mL of 0.09543 N NaOH.

7. Complete and balance an equation for the following redox reaction in acid solution:

$$As_2O_3 + CrO_4^{2-} \rightarrow H_3AsO_4 + Cr^{3+}$$

8. Complete and balance an equation for the following redox reaction in acid solution:

$$Pb + PbO_2 + H_2SO_4 \rightarrow PbSO_4$$

9. Complete and balance an equation for the following redox reaction in acid solution:

$$H_2Cr_2O_7 + Zn \rightarrow Cr^{3+} + Zn^{2+}$$

10. Name the following compounds using the Stock system:

   (a) ClO_2 _____

   (b) Cl_2O_5 _____

   (c) ClO_3 _____

   (d) Cl_2O_7 _____

11. Predict formulas for the possible compounds between the pairs of elements:

   (a) tin and chlorine _____

   (b) selenium and bromine _____

12. Calculate the normality of a base if 50.00 mL of 0.09942 N HCl is neutralized by 56.78 mL of the base.

13. Circle the following substances that are good oxidizing agents.

$KMnO_4$ H_2 Mg O_2 $K_2Cr_2O_7$ KI

14. Complete and balance the following equation:

$$Zn + Cu(NO_3)_2 \rightarrow Zn(NO_3)_2 + Cu$$

Complete the following sentences using the balanced equation:

_____ is oxidized.

_____ is reduced.

_____ is the oxidizing agent.

_____ is the reducing agent.

15. Complete and balance the following equation:

$$P_4 + NaOH \rightarrow PH_3 + NaH_2PO_2$$

Complete the following sentences using the balanced equation:

_____ is oxidized.

_____ is reduced.

_____ is the oxidizing agent.

_____ is the reducing agent.

16. Calculate the equivalent mass of an unknown acid if 3.500 grams of the acid require 38.64 mL of 0.01978 N NaOH for neutralization.

Chapter 16: Experiment at Home

Measuring the Concentration of Oxygen in Air

When iron rusts, it reacts with oxygen in the air to form iron(III) oxide. In this experiment you will allow iron to rust in air and in doing so use up the oxygen in the air. By measuring the volume of the air before and after the experiment, you will be able to tell how much oxygen was originally in the air. The actual reaction is a redox reaction, but you will also need to remember the gas laws and stoichiometry in order to calculate the result from these data.

Materials: glass casserole dish
glass baby bottle
iron filings or steel wool cut into small pieces
rubber band (to fit around baby bottle)
measuring cup calibrated in milliliters
thermometer
barometer

Method:

(1) Measure the volume of the baby bottle by filling it completely to the top and then carefully pouring the contents into the measuring cup.

(2) Measure the temperature of the water in degrees Celsius. Record the vapor pressure of water at this temperature from Table 12.3 in your textbook. Record the barometric pressure from a barometer or by using the weather service data from the radio or telephone.

(3) Put the rubber band around the baby bottle. Place the iron filings or cut steel wool into the wet bottle and allow them to stick to the sides.

(4) Invert the bottle into the casserole dish full of water. Be careful that no air bubbles escape. Some water will rise into the bottle. Allow the apparatus to stand undisturbed for about one week.

(5) After one week, water will have replaced the reacted oxygen in the bottle. Rust will also appear on the iron. Move the bottle so that the level of the water is the same inside as it is outside. You will have to raise the bottle carefully. When you have this level, move the rubber band to indicate the level.

(6) Turn the bottle over and add water to the level of the rubber band. Measure the volume of the water by pouring it carefully into the measuring cup. This will be the amount of air that was *not* oxygen.

(7) Again record the temperature and barometric pressure as you did in step 2 above. Record the vapor pressure of water at this temperature from Table 12.3.

(8) Complete the data sheet and calculate the percentage of oxygen in air on the next page.

Measuring the Concentration of Oxygen in Air

Experimental Data -- Start of Experiment   Date:

(1) Volume of baby bottle: _____ mL

(2) Temperature: _____ kelvins (T_1)

(3) Vapor pressure of water: _____ mm Hg

(4) Barometric pressure: _____ mm Hg

(5) Pressure of dry air [(4) - (3)] _____ mm Hg (P_1)

(6) Volume of dry air [(5)/(4) x (1)] _____ mL (V_1)

(7) Using (2), (5), and (6) as T_1, P_1, and V_1, calculate the volume of the air at STP.

$$V_2 = [P_1 V_1 (273)] / [(760 \text{ mm Hg}) T_1] = \underline{} \text{ mL}$$

Experimental Data -- End of Experiment   Date:

(8) Volume of water: _____ mL

(9) Temperature: _____ kelvins (T_2)

(10) Vapor pressure of water: _____ mm Hg

(11) Barometric pressure: _____ mm Hg

(12) Pressure of dry air [(11) - (10)] _____ mm Hg (P_2)

(13) Volume of dry air [(12)/(11) x (1)] _____ mL (V_2)

(14) Using (9), (12), and (13) as T_2, P_2, and V_2, calculate the volume of air left after oxygen removal at STP.

$$V = [P_2 V_2 (273)] / [(760 \text{ mm Hg}) T_2] = \underline{} \text{ mL}$$

(15) The volume of oxygen is calculated by subtracting (14) from (7). _____ mL

(16) The percentage of oxygen is [(15)/(7) x 100] _____ %

(17) Air is approximately 21% oxygen. What is your percentage error?

Chapter 17: Reaction Rates and Chemical Equilibrium

Many chemists refer to equilibrium as a *steady state* because there is no net change of anything. The important word is net. In any chemical system at equilibrium you are dealing with a *dynamic* system. The rate forward is equal to the rate backward. The reaction does not shut down when equilibrium is reached; it simply maintains constant concentrations of reactants and products. For this reason it is easy to understand that when some stress is exerted on the system it will alter either the forward or the reverse reaction in order to re-establish the equilibrium.

Section 17.1: Rates of Reaction

☐ Read Section 17.1.

☐ List the six factors that can affect the rate of a chemical reaction.

1. _____
2. _____
3. _____
4. _____
5. _____
6. _____

These can be found in the first paragraph of Section 17.1.

☐ Make sure to read the Examples and work the Practice Problems in the text as you read this section. Check your answers carefully.

☐ Work the Problems for this section at the end of the chapter.

Section 17.2: The Condition of Equilibrium

☐ Read Section 17.2.

☐ Write the two equations used as examples in this section.

_____

_____

☐ Write the two equations as one reaction using a double arrow. Does it matter which compound you place on the right or the left?
Write it both ways.

_____

_____

☐ In each of the equations above circle the reactants and place a box around the products. The reactants are always the substances on the *left* side of the reaction. The products are always the substances on the *right* side of the equation.

☐ Define each of the following expressions in your own words:

   proceed to the right: _____

_____

   proceed to the left: _____

_____

☐ Make sure to read the Examples and work the Practice Problems in the text as you read this section. Check your answers carefully.

☐ Work the Problems for this section at the end of the chapter.

Section 17.3: LeChâtelier's Principle

☐ Read Section 17.3.

One way of looking at a system in equilibrium is in terms of "pushing" and "pulling" the equation. Whenever you add something to a system at equilibrium it will have the effect of "pushing" the equation to the other side. If something is removed it will "pull" the equation back toward that same side.

Pressure changes are somewhat different, but also follow the rule that the system will compensate for added stress by reducing the stress. If pressure is added, the system will re-establish equilibrium by lowering the number of moles of gas present. If pressure is reduced, then the reaction will shift to produce more moles of gas.

☐ Make sure to read the Examples and work the Practice Problems in the text as you read this section. Check your answers carefully.

☐ Work the Problems for this section at the end of the chapter.

Section 17.4: Equilibrium Constants

☐ Read Section 17.4.

Note carefully how the equilibrium expression is written from the equation. This pattern must be used whenever you write an equilibrium expression. The brackets used in the equilibrium expression indicate molarity (moles/liter).

Important information can be obtained by just looking at the value of the equilibrium constant. Remember that the products appear in the numerator and the reactants are in the denominator of the fraction.

$$\frac{[Products]}{[Reactants]}$$

When the value of K is greater than 10^4

[products] >> [reactants]   The equilibrium lies to the right.

When the value of K is less that 10^{-4}

[products] << [reactants]   The equilibrium lies to the left.

☐ Read carefully the five rules listed in this section.

☐ Make sure to read the Examples and work the Practice Problems in the text as you read this section. Check your answers carefully.

☐ Work the Problems for this section at the end of the chapter.

Chapter 17: Finishing Up

☐ Carefully read the summary section at the end of the chapter. Do you understand each paragraph? Do you know the terms used? If not, review the section indicated at the end of the paragraph.

☐ The section titled "Items for Special Attention" gives you some pointers to help you learn the material correctly and also can keep you from making some of the more common mistakes. Read these carefully.

☐ Work the "Self-Tutorial Problems." Check each answer as you complete the problem. If you have trouble with one of these, be sure to ask your instructor for help.

☐ Look over the list of "Key Terms" at the beginning of the chapter. If you do not recognize a term or are unsure of how it was used in the chapter, go back to that section and reread it.

☐ Use your lecture notes and the text to find the important topics covered in class. Make flash cards so that you can study these carefully before the exam.

☐ Work each of the problems at the end of the chapter in your text. Be sure to check your answers. If the answer is not given for that problem, work another that is similar and check the answer. If you are incorrect, check your work and review the chapter to see if you can answer the question correctly. If you cannot, get help from your instructor.

☐ Work the sample exam questions on the next few pages.

☐ Check your answers when you have completed all of them.

☐ Make up your own exam and exchange it with a fellow student or use it in your study group.

Chapter 17: Sample Exam Questions

1. Balance the following equation:

$$4NO_2(g) + 6H_2O(g) \rightleftharpoons 7O_2(g) + 4NH_3(g)$$

(a) Write the equilibrium expression for this reaction.

(b) Given that this is an endothermic reaction, predict the direction of equilibrium shift (L or R) for each of the following:

_____ increasing the temperature.

_____ adding water.

_____ removing NO_2.

_____ adding ammonia.

_____ adding a catalyst.

2. Balance the following equation:

$$F_2(g) + PH_3(g) \rightleftharpoons HF(g) + PF_3(g)$$

(a) Write the equilibrium expression for this reaction.

(b) Predict the direction of equilibrium shift (L or R) for each of the following:

_____ increasing the pressure.

_____ adding F_2.

_____ removing PF_3.

_____ adding HF.

_____ adding a catalyst.

3. Write the chemical equations corresponding to the following equilibrium expressions:

(a) $K = \dfrac{[CO][Cl_2]}{[COCl_2]}$

_____

(b) $K = \dfrac{[H_2][Br_2]}{[HBr]^2}$

_____

(c) $K = \dfrac{[O_2]^9[C_3H_6]^2}{[CO_2]^6[H_2O]^6}$

_____

4. When a mixture of H_2, CO_2, H_2O, and CO at 987°C reached equilibrium, the concentrations were:

$[H_2] = 2.63$ M $\qquad$ $[H_2O] = 0.808$ M
$[CO_2] = 0.0962$ M $\qquad$ $[CO] = 0.473$ M

Calculate the equilibrium constant at 987°C for the reaction:

$$H_2 + CO_2 \rightleftharpoons H_2O + CO$$

5. A reaction has an equilibrium constant of 1.78×10^{-5} at a temperature of 586°C. The same reaction has an equilibrium constant of 6.54×10^{-2} at a temperature of 378°C. To increase the amount of products, at which of these temperatures should the reaction be run?

6. Write the equilibrium expression for the following equation:

$$C_2H_6 + 2\ Cl_2 \rightleftharpoons C_2H_4Cl_2 + 2\ HCl$$

7. For the following equation, predict the direction of the equilibrium shift caused by the indicated change:

$$2\ NO_2(g) \rightleftharpoons N_2O_4(g)$$

(a) addition of NO_2 _____

(b) addition of N_2O_4 _____

(c) increased pressure _____

8. For the following endothermic reaction of gases, predict the direction of the equilibrium shift caused by the indicated change:

$$4\ NO_2 + 6\ H_2O \rightleftharpoons 7\ O_2 + 4\ NH_3$$

(a) addition of NO_2 _____

(b) removal of NH_3 _____

(c) removal of heat _____

(d) decrease of pressure _____

9. Calculate the equilibrium constant for the reaction of hydrogen gas with bromine gas to produce hydrogen bromide gas, given the following equilibrium concentrations:

$$[H_2] = 0.37 \text{ mol/L} \qquad [Br_2] = 0.70 \text{ mol/L} \qquad [HBr] = 3.6 \text{ mol/L}$$

10.. The following reaction has an equilibrium constant of 10:

$$C_2H_4 + Cl_2 \rightleftharpoons C_2H_4Cl_2$$

What is the equilibrium concentration of $C_2H_4Cl_2$, when $[C_2H_4] = 0.50$ mol/L and $[Cl_2] = 0.30$ mol/L?

11. If the reaction in Problem 10 has an equilibrium constant of 25 at a different temperature, what is the equilibrium concentration of $C_2H_4Cl_2$, when $[C_2H_4] = 0.15$ mol/L and $[Cl_2] = 0.30$ mol/L?

12. The concentration of reactants was measured at several points during a reaction. From the data below, determine the reaction rate (change in molarity per second, M/s) for the following reaction:

$$N_2O_4 \rightleftharpoons 2 NO_2$$

Time in seconds	Concentration of N_2O_4
0	0.40 M
20	0.36 M
40	0.32 M
60	0.28 M

Chapter 17: Experiment at Home

Photochromic and Thermochromic Reactions

Many chemical reactions are photochromic and thermochromic. These are reactions that change color (-chromic) driven by light (photo-) or heat (thermo-). Some of the most dramatic of these reactions involve the equilibria of different metal salts. Silver salts have long been used on photographic plates. There are also many organic compounds that change color when illuminated or heated. Because many of these chemical reactions are reversible, you will be able to demonstrate LeChâtelier's principle using light and heat.

Materials: 1 teaspoon of cornstarch
several drops of tincture of iodine or Betadine®
1 small glass or vial
1 pan of boiling water
1 vitamin C tablet

Method: Fill the vial half full of water. Mix in a pinch of cornstarch and 6 to 10 drops of iodine, enough to give the solution a deep blue-black color. Heat the vial by holding it in the boiling water. After several seconds the heat should cause the color to disappear. Removing the vial from the heat should cause the color to shift back to the blue-black. This can be repeated several times, although you may have to add some more iodine occasionally. The addition of vitamin C will irreversibly remove the color.

Theory: The mixture of starch and iodine forms a loose complex that absorbs light. Heating disrupts this complex and therefore causes the solution to become clear. Recooling drives the equilibrium back in favor of the light absorbing complex. This can be repeated as long as some iodine is still present in solution. Some of the iodine is removed by the slow oxidation of starch. By adding the vitamin C, the iodine is completely reduced to iodide and the equilibrium no longer exists.

Chapter 18: Acid-Base Theory

Now that you have learned about equilibrium you will be able to understand the equilibrium of weak acids and bases. The weak acids and bases are the ones you encounter everyday. Acetic acid (vinegar), citric acid (orange juice), and ascorbic acid (vitamin C) are common weak acids found in food. Ammonia is a common weak base found in household cleaning solutions.

Throughout the acid-base theory of this chapter you will be dealing with *proton transfer reactions*. This just means that whenever you look at an acid base reaction you should look for the position of the H^+ ion.

When water accepts a proton, it is written as H_3O^+. This is called the *hydronium ion*. The OH^- ion is called the *hydroxide ion*. The interaction of these two ions in solutions is the subject of this chapter.

Section 18.1: The Brønsted Theory

☐ Read Section 18.1.

☐ Look at the examples of conjugate acid-base pairs in your text. Note which species on each side of the reaction has the **H**.

$$\textbf{H}Cl(g) + H_2O(l) \rightarrow \textbf{H}\text{-}H_2O^+(aq) + Cl^-(aq)$$

The hydrogen is donated by the HCl and the H_2O accepts it.

The H_2O becomes H_3O^+ and the HCl becomes Cl^-.

$$HCl(g) + H_2O(l) \rightarrow H_3O^+(aq) + Cl^-(aq)$$

Labeling these as conjugate acid-base pairs is simply looking at the reactant side of the reaction and determining which compound loses (donates) the proton and which compound has accepted it on the product side of the reaction.

☐ Label the following reaction to show the conjugate acid-base pairs.

$$HCl(g) + H_2O(l) \rightarrow H_3O^+(aq) + Cl^-(aq)$$

Water is described as amphiprotic, that is, being able to donate or accept a proton. In the following reaction, label the conjugate acid-base pairs.

$$H_2O + H_2O \rightleftharpoons H_3O^+ + OH^-$$

```
     acid                                          conjugate base
      ↓                                                  ↓
     H₂O      +      H₂O     ⇌      H₃O⁺       +       OH⁻
                      ↑               ↑
                     base         conjugate acid
```

☐ Note Table 18.1. Remember these relationships.

In Chapter 8 you learned to recognize strong acids and strong bases. The table below lists these. Before you continue in this chapter you need to memorize them. If an acid or base is not on this list, you may assume that it is weak.

Strong acids	HCl HBr HI HClO$_3$ HClO$_4$ HNO$_3$ H$_2$SO$_4$
Strong bases	LiOH NaOH KOH RbOH CsOH Ba(OH)$_2$

☐ Make sure to read the Examples and work the Practice Problems in the text as you read this section. Check your answers carefully.

☐ Work the Problems for this section at the end of the chapter.

Section 18.2: Ionization Constants

The degree to which an acid or base dissociates, producing ions in solution, determines its strength. As the dissociation increases, the acid or base becomes stonger. You can use ionization constants to determine the stronger of two weak acids. This section gives you two types of calculations based on ion dissociation constants. First, you should be able to tell the concentration of all species in solution given the undissociated concentration of a weak acid or base. Second, given the concentration of the hydrogen or hydroxide ion, you should be able to calculate the dissociation constant.

☐ Read Section 18.2.

☐ Make sure to read the Examples and work the Practice Problems in the text as you read this section. Check your answers carefully.

☐ Work the Problems for this section at the end of the chapter.

Section 18.3: Autoionization of Water

In this section you will learn a very useful concentration unit, pH. This unit is based on the value of the water ionization constant, K_w.

☐ Read Section 18.3.

Memorize the following two pieces of information:

$$K_w = 1.00 \times 10^{-14}$$

$$pH = -\log[H_3O^+]$$

Sometimes rearranging the pH equation is useful to its understanding. Solving for $[H_3O^+]$,

$$[H_3O^+] = 10^{-pH}$$

Using this equation you can tell quickly that if a solution has an $[H_3O^+]$ of 10^{-4}, then the pH of the solution is 4.

Another useful tool for beginning students is to look at the $[H_3O^+]$ values in non-scientific notation. Look over the following table carefully.

$[H_3O^+]$ (M)	$[H_3O^+]$	pH	Type of solution
1.0	10^0	0	
0.1	10^{-1}	1	strong acids
0.01	10^{-2}	2	
0.001	10^{-3}	3	
0.0001	10^{-4}	4	
0.00001	10^{-5}	5	weak acids
0.000001	10^{-6}	6	
0.0000001	10^{-7}	7	neutral
0.00000001	10^{-8}	8	
0.000000001	10^{-9}	9	
0.0000000001	10^{-10}	10	weak bases
0.00000000001	10^{-11}	11	
0.000000000001	10^{-12}	12	strong bases
0.0000000000001	10^{-13}	13	
0.00000000000001	10^{-14}	14	

Note that as the pH gets larger, the $[H_3O^+]$ gets smaller. This is often a source of confusion. Look over the powers of 10 in column two and compare them to the values in column one to see the relationship more clearly.

Using the K_w, you can see what happens to the $[OH^-]$ as the $[H_3O^+]$ decreases. The following table will help you understand.

$[H_3O^+]$ (M)	$[H_3O^+]$	pH	$[OH^-]$	$[OH^-]$ (M)
1.0	10^0	0	10^{-14}	0.00000000000001
0.1	10^{-1}	1	10^{-13}	0.0000000000001
0.01	10^{-2}	2	10^{-12}	0.000000000001
0.001	10^{-3}	3	10^{-11}	0.00000000001
0.0001	10^{-4}	4	10^{-10}	0.0000000001
0.00001	10^{-5}	5	10^{-9}	0.000000001
0.000001	10^{-6}	6	10^{-8}	0.00000001
0.0000001	10^{-7}	7	10^{-7}	0.0000001
0.00000001	10^{-8}	8	10^{-6}	0.000001
0.000000001	10^{-9}	9	10^{-5}	0.00001
0.0000000001	10^{-10}	10	10^{-4}	0.0001
0.00000000001	10^{-11}	11	10^{-3}	0.001
0.000000000001	10^{-12}	12	10^{-2}	0.01
0.0000000000001	10^{-13}	13	10^{-1}	0.1
0.00000000000001	10^{-14}	14	10^0	1.0

To go one step further in understanding pH, look carefully at Example 18.11. You are asked to find the pH of a solution with $[H_3O^+] = 2.89 \times 10^{-1}$ M, $[H_3O^+] = 2.89 \times 10^{-7}$ M, and $[H_3O^+] = 2.89 \times 10^{-13}$ M.

2.89×10^{-1} is between 10^0 and 10^{-1}. In other words, the value is greater than 10^{-1} and less than 10^0. The negative logarithm of this number should therefore be between 0 and 1. The answer of 0.539 is within this range.

2.89×10^{-7} is between 10^{-6} and 10^{-7}. The pH should be between 6 and 7. The answer of 6.539 is reasonable.

2.89×10^{-13} is between 10^{-12} and 10^{-13}. The pH should be between 12 and 13. The answer of 12.539 is reasonable.

Practice using your calculator so that you will always get reasonable answers. Use the above steps to check your calculations.

☐ Make sure to read the Examples and work the Practice Problems in the text as you read this section. Check your answers carefully.

☐ Work the Problems for this section at the end of the chapter.

Section 18.4: Buffer Solutions

Buffer solutions are very important to us biochemically. They protect organisms from extremes of pH. In order to prepare a buffer solution with a desired pH, an appropriate weak acid-conjugate base pair is selected. A useful equation for this is the Henderson-Hasselbach equation. This is simply the logarithmic form of the equilibrium expression.

$$pH = pK_a + \log([salt]/[acid])$$

The letter "p" is a mathematical function for the negative logarithm of a number. The salt concentration is the molarity of the conjugate base, the acid concentration is the molarity of the weak acid. pK_a values can be found in tables. This expression shows that when the [salt] = [acid] the logarithm of this ratio is zero and the resulting pH of the buffer is the same as the pK_a of the weak acid.

☐ Read Section 18.4.

☐ Make sure to read the Examples and work the Practice Problems in the text as you read this section. Check your answers carefully.

☐ Work the Problems for this section at the end of the chapter.

Chapter 18: Finishing Up

☐ Carefully read the summary section at the end of the chapter. Do you understand each paragraph? Do you know the terms used? If not, review the section indicated at the end of the paragraph.

☐ The section titled "Items for Special Attention" gives you some pointers to help you learn the material correctly and also can keep you from making some of the more common mistakes. Read these carefully.

☐ Work the "Self-Tutorial Problems." Check each answer as you complete the problem. If you have trouble with one of these, be sure to ask your instructor for help.

☐ Look over the list of "Key Terms" at the beginning of the chapter. If you do not recognize a term or are unsure of how it was used in the chapter, go back to that section and reread it.

☐ Use your lecture notes and the text to find the important topics covered in class. Make flash cards so that you can study these carefully before the exam.

☐ Work each of the problems at the end of the chapter in your text. Be sure to check your answers. If the answer is not given for that problem, work another that is similar and check the answer. If you are incorrect, check your work and review the chapter to see if you can answer the question correctly. If you cannot, get help from your instructor.

☐ Work the sample exam questions on the next few pages.

☐ Check your answers when you have completed all of them.

☐ Make up your own exam and exchange it with a fellow student or use it in your study group.

Chapter 18: Sample Exam Questions

1. Calculate the pH of the following solutions:

    (a) $[H_3O^+] = 1.45 \times 10^{-11}$ M _____

    (b) $[OH^-] = 5.38 \times 10^{-5}$ M _____

2. Calculate the $[H_3O^+]$ for each of the following:

    (a) pH = 8.76 _____

    (b) pH = 0.12 _____

    (c) $[OH^-] = 9.27 \times 10^{-9}$ M _____

3. Calculate the pH of 1.50 L of a solution originally containing 6.82×10^{-5} M NaOH after 1.25 L of 9.37×10^{-7} M HCl has been added to it.

4. Calculate the pH of the following solutions:

    (a) $[H_3O^+] = 2.69 \times 10^{-8}$ M _____

    (b) $[OH^-] = 9.51 \times 10^{-12}$ M _____

5. Calculate the $[H_3O^+]$ for each of the following:

    (a) pH = 10.34 _____

    (b) pH = 1.12 _____

    (c) $[OH^-] = 9.27 \times 10^{-12}$ M _____

6. Calculate the pH of 150 mL of a solution originally containing 5.96×10^{-4} M NaOH after 125 mL of 9.37×10^{-3} M HCl has been added to it.

7. Write the formula for the conjugate *base* of each of the following:

   (a) HSO_4^- _____

   (b) $H_2PO_4^-$ _____

   (c) HCO_3^- _____

   (d) H_2O _____

8. Write the formula for the conjugate *acid* of each of the following:

   (a) H_2O _____

   (b) $H_2PO_4^-$ _____

   (c) HSO_4^- _____

   (d) HCO_3^- _____

9. Calculate the pH of each of the following solutions, given the $[H_3O^+]$:

   (a) 2.35×10^{-10} M _____

   (b) 5.28×10^{-5} M _____

   (c) 8.97×10^{-8} M _____

   (d) 3.15×10^{-7} M _____

10. Calculate the $[H_3O^+]$ for each of the following pH values:

   (a) 1.678   _____

   (b) 6.982   _____

   (c) 11.453   _____

   (d) –0.217   _____

11. Calculate the hydronium ion concentration of 0.200 M acetic acid. The K_a for acetic acid is 1.8×10^{-5}. Acetic acid dissolves in water by the following reaction:

$$HC_2H_3O_2 + H_2O \rightleftharpoons C_2H_3O_2^- + H_3O^+$$

12. Calculate the percent ionization of a 0.200 M methylamine solution.
    $$CH_3NH_2 + H_2O \rightleftharpoons CH_3NH_3^+ + OH^- \qquad (K_b = 4.4 \times 10^{-4})$$

13. Calculate the pH of each of the following solutions, given the $[OH^-]$:

   (a) 2.35×10^{-10} M   _____

   (b) 5.28×10^{-5} M   _____

   (c) 8.97×10^{-8} M   _____

   (d) 3.15×10^{-7} M   _____

14. A 0.150 M solution of a weak base, B, has a pH of 8.34. Calculate the value of K_b for this base.

15. Which of the following combinations will form a buffer solution in water?

   (a) 0.250 M $HC_2H_3O_2$ and 0.250 M NaOH

   (b) 0.150 M $HC_2H_3O_2$ and 0.150 M $NaC_2H_3O_2$

   (c) 0.050 M $NaC_2H_3O_2$ and 0.050 M HCl

   (d) 0.125 M Na_2HPO_4 and 0.125 M NaH_2PO_4

16. Determine whether each of the following 0.100 M solutions would be acidic, basic, or neutral:

   (a) K_3PO_4 _____

   (b) Li_2CO_3 _____

   (c) $(NH_4)_2SO_4$ _____

   (d) $CuCl_2$ _____

   (e) $NaNO_3$ _____

Chapter 18: Experiment at Home

Red Cabbage Indicator Solution

Materials: small head of red cabbage
cooking pan
small white styrofoam cups
water
straws
household chemicals

Cut about one-fourth of the cabbage into strips about one-half inch wide. Place the cabbage into the pan and add about one cup of water. Bring to a boil and then allow to cool. Using a straw as a pipet, transfer a different household solution into each of the cups. Hold the straw in the solution until about one inch of solution is in the straw.

DO NOT USE YOUR MOUTH TO DRAW UP THE SOLUTIONS!!!

Place your finger over the top end of the straw and the solution will stay in the straw until you are ready to deliver it to the styrofoam cup. Use a clean straw for each product tested. Solids will need to be dissolved in a small amount of water for testing. Be sure to label each cup so that you will know what you put in it.

Add the same amount of cabbage juice in the same manner as above to each cup.

The variation in color caused by the pH of the various chemicals is very impressive. By testing some of the substances in Table 18.3 you can set up a pH scale. Try to arrange the colors in order of increasing pH.

When you are finished with the experiment, pour your samples down the sink and flush with lots of water. The cabbage juice will keep, in a closed glass container, in the refrigerator for several weeks.

Other items you may wish to try include: turmeric, tea, rhubarb, grape juice, purple hollyhocks, violets, and cherries. Just crush, squeeze, or mash and place in just enough water so that you can see the color of the solution. Test with known household acids and bases to see the range of colors.

Chapter 19: Organic Chemistry

If you are using this course to prepare for a health career, you will want to pay close attention to the material in this chapter. Much of the information will be used in nutrition and physiology courses you may be required to take. The subjects of this chapter relate very closely to the material you have already studied. Some of the subjects you should review before starting this chapter include:

☐ **covalent bonding** You should remember the characteristics of the bonds formed by sharing electrons (Section 5.4). Know how to determine whether two atoms are bonded by a single, double, or triple bond.

☐ **bond orders of covalent molecules** These can be found in Table 19.3. You worked with most of these in Chapter 5 when you learned to draw electron dot structures.

☐ **types of reactions** Combustion reactions were introduced in Section 8.3. You will need to be able to write these for both the complete and incomplete combustion of hydrocarbons.

☐ **hydrogen bonding** Section 13.6 describes the requirements for hydrogen bonding. Hydrogen bonding is important in organic compounds containing nitrogen, oxygen, or both. As you read Chapter 19, classify the different types of compounds by whether they can participate in hydrogen bonding. The presence of hydrogen bonding explains why some compounds boil at much higher temperatures than would be expected from their molecular weights. The ability of a molecule to hydrogen bond with water molecules will tell you about the solubility of the substance in water.

☐ **acid and base reactions** Weak acids and ammonia derivatives of Section 8.4 and Chapter 19 are fundamental to the understanding of amino acids.

Section 19.1: Hydrocarbons

Most students find that flash cards are the best way to learn the properties of the various types of organic compounds. You will want to identify the properties of each class of compounds on one side and the name of the class on the other.

The following tables of Chapter 19 must be memorized:

☐ **Table 19.1** In addition to the material in this table you should also memorize the prefixes for the carbon chain lengths. For example, when you see *meth-*, you should think one carbon, *eth-* means two carbons, and so forth.

☐ Fill in the following table using the names from Table 19.1.

three carbon chain _____

six carbon chain _____

five carbon chain _____

two carbon chain _____

seven carbon chain _____

one carbon chain _____

nine carbon chain _____

four carbon chain _____

eight carbon chain _____

ten carbon chain _____

☐ Be sure to check your answers with the names in Table 19.1.

☐ Table 19.2 The names ethylene and propylene are used so commonly that you should learn them in addition to the systematic names.

☐ Table 19.3 The bond orders are the same ones you learned when doing electron dot structures in Chapter 5. Knowing the bond orders is important when you are interpreting **line formulas**.

☐ Table 19.4 This table summarizes the first three sections of the chapter. It is a useful method for organizing your notes on the classes of compounds.

☐ Read Section 19.1. Take careful notes on each type of hydrocarbon. For the following classes which you will study, the reactions of organic compounds are generally divided into three main categories: substitution, addition, and elimination. As you read about each class of organic compound, keep track of the type of reaction associated with it.

Alkanes, alkenes, and alkynes are often referred to as *aliphatic* hydrocarbons. *Aromatic* hydrocarbons are the ones that contain a benzene ring.

Aromatic hydrocarbons have special properties because of the benzene ring, called a *conjugated* ring structure. All of the bonds in benzene average out to be the same. Because the double bonds and single bonds in the ring are constantly changing places, only bonds that are a hybrid (mixed) formation are detected. This *resonance* helps to explain the stability of benzene rings and why they do not undergo addition reactions.

Explain the following terms in your own words:

saturated hydrocarbon: _____

_____

unsaturated hydrocarbon: _____

_____

aromatic hydrocarbon: _____

_____

line formula: _____

_____

isomer: _____

_____

derivative: _____

_____

structural formula: _____

_____

tetrahedron: _____

_____

alkane: _____

_____

alkene: _____

_____

☐ Make sure to read the Examples and work the Practice Problems in the text as you read this section. Check your answers carefully.

☐ Work the Problems for this section at the end of the chapter.

Section 19.2: Isomerism

☐ Read Section 19.2.

☐ Make sure to read the Examples and work the Practice Problems in the text as you read this section. Check your answers carefully.

☐ Work the Problems for this section at the end of the chapter.

☐ Heptane has nine isomers. Try to write them on a piece of paper. When you have completed this, check with the answer below. The answers are given below with line formulas. You may need to write these as structural formulas to see the isomers more clearly. The names of the compounds are also given.

$CH_3(CH_2)_5CH_3$ heptane

$(CH_3)_2CH(CH_2)_3CH_3$ 2-methylhexane

$CH_3CH_2CH(CH_3)(CH_2)_2CH_3$ 3-methylhexane

$(CH_3)_3C(CH_2)_2CH_3$ 2,2-dimethylpentane

CH$_3$CH$_2$C(CH$_3$)$_2$CH$_2$CH$_3$	3,3-dimethylpentane
(CH$_3$)$_2$CHCH(CH$_3$)CH$_2$CH$_3$	2,3-dimethylpentane
(CH$_3$)$_2$CHCH$_2$CH(CH$_3$)$_2$	2,4-dimethylpentane
(CH$_3$)$_3$CCH(CH$_3$)$_2$	2,2,3-trimethylbutane
(CH$_3$CH$_2$)$_3$CH	3-ethylpentane

Section 19.3: Other Classes of Organic Compounds

☐ Read Section 19.3.

☐ Define the following terms used in Section 19.3.

radical: _____

functional group: _____

carbonyl group: _____

ionizable hydrogen atom: _____

halide: _____

dialcohol (diol) _____

total bond order: _____

☐ Organize the information (reactions, nomenclature, common names) on each of the functional groups into a pattern which you can easily memorize. Be sure to incorporate your lecture notes in case your instructor gives you additional information on each of these topics.

☐ Make sure to read the Examples and work the Practice Problems in the text as you read this section. Check your answers carefully.

☐ Work the Problems for this section at the end of the chapter.

Section 19.4: Polymers

Probably the most important class of compounds of this decade is that of polymers. The world of plastic around us attests to the flexibility of these compounds. Our own bodies are made of biopolymers such as proteins, enzymes, RNA, and DNA. The backbone of a polymer is a *repeating unit*. The properties of the polymer are determined by *side groups* along the chain.

☐ Read Section 19.4.

☐ Using the information in Table 19.5, see if you can draw several repeating units of the polymers listed.

Below are several examples of polymer formation. Note that the *repeating unit* is in a box.

Polypropylene

$$-\underset{\underset{H}{|}}{\overset{\overset{H}{|}}{C}} - \underset{\underset{H}{|}}{\overset{\overset{CH_3}{|}}{C}} - \boxed{\underset{\underset{H}{|}}{\overset{\overset{H}{|}}{C}} - \underset{\underset{H}{|}}{\overset{\overset{CH_3}{|}}{C}}} - \underset{\underset{H}{|}}{\overset{\overset{H}{|}}{C}} - \underset{\underset{H}{|}}{\overset{\overset{CH_3}{|}}{C}} -$$

Polyvinyl chloride:

$$-\underset{\underset{H}{|}}{\overset{\overset{H}{|}}{C}} - \underset{\underset{H}{|}}{\overset{\overset{Cl}{|}}{C}} - \boxed{\underset{\underset{H}{|}}{\overset{\overset{H}{|}}{C}} - \underset{\underset{H}{|}}{\overset{\overset{Cl}{|}}{C}}} - \underset{\underset{H}{|}}{\overset{\overset{H}{|}}{C}} - \underset{\underset{H}{|}}{\overset{\overset{Cl}{|}}{C}} -$$

Polystyrene

$$-\underset{\underset{H}{|}}{\overset{\overset{H}{|}}{C}} - \underset{\underset{H}{|}}{\overset{\overset{C_6H_5}{|}}{C}} - \boxed{\underset{\underset{H}{|}}{\overset{\overset{H}{|}}{C}} - \underset{\underset{H}{|}}{\overset{\overset{C_6H_5}{|}}{C}}} - \underset{\underset{H}{|}}{\overset{\overset{H}{|}}{C}} - \underset{\underset{H}{|}}{\overset{\overset{C_6H_5}{|}}{C}} -$$

Polytetrafluoroethylene (Teflon™)

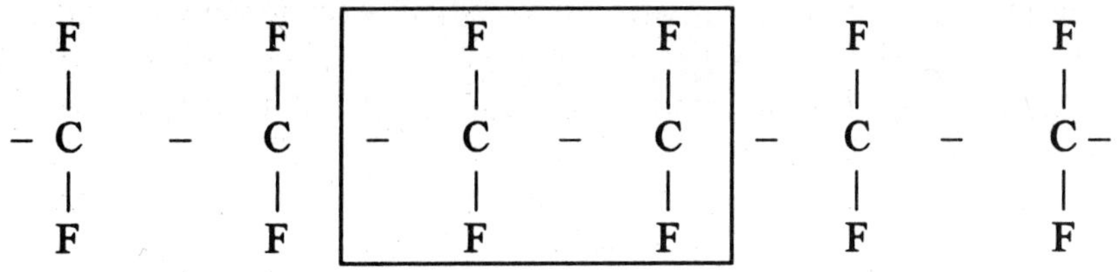

$$-\overset{F}{\underset{F}{C}}- \overset{F}{\underset{F}{C}} - \boxed{-\overset{F}{\underset{F}{C}}- \overset{F}{\underset{F}{C}} -} -\overset{F}{\underset{F}{C}}- \overset{F}{\underset{F}{C}}-$$

☐ Make sure to read the Examples and work the Practice Problems in the text as you read this section. Check your answers carefully.

☐ Work the Problems for this section at the end of the chapter.

Section 19.5: Foods

☐ Read Section 19.5.

☐ Fill in the following table with definitions for the words from this section:

fat: _____

_____

glycerol (glycerine): _____

_____

triester: _____

_____

soap: _____

_____

hydrogenate: _____

_____

carbohydrate: _____

_____

sugar: _____

_____

monosaccharide: _____

_____

disaccharide: _____

_____

polysaccharide: _____

_____

☐ Make sure to read the Examples and work the Practice Problems in the text as you read this section. Check your answers carefully.

☐ Work the Problems for this section at the end of the chapter.

Chapter 19: Finishing Up

☐ Carefully read the summary section at the end of the chapter. Do you understand each paragraph? Do you know the terms used? If not, review the section indicated at the end of the paragraph.

☐ The section titled "Items for Special Attention" gives you some pointers to help you learn the material correctly and also can keep you from making some of the more common mistakes. Read these carefully.

☐ Work the "Self-Tutorial Problems." Check each answer as you complete the problem. If you have trouble with one of these, be sure to ask your instructor for help.

☐ Look over the list of "Key Terms" at the beginning of the chapter. If you do not recognize a term or are unsure of how it was used in the chapter, go back to that section and reread it.

☐ Use your lecture notes and the text to find the important topics covered in class. Make flash cards so that you can study these carefully before the exam.

☐ Work each of the problems at the end of the chapter in your text. Be sure to check your answers. If the answer is not given for that problem, work another that is similar and check the answer. If you are incorrect, check your work and review the chapter to see if you can answer the question correctly. If you cannot, get help from your instructor.

☐ Work the sample exam questions on the next few pages.

☐ Check your answers when you have completed all of them.

☐ Make up your own exam and exchange it with a fellow student or use it in your study group.

Chapter 19: Sample Exam Questions

1. Identify the class of each of the following compounds:

    (a) $CH_3CH_2CH_2OH$ _____

    (b) CH_3CHO _____

    (c) $CH_3CH_2OCH_2CH_2CH_3$ _____

    (d) $CH_3CH_2CH=CH_2$ _____

    (e) $CH_3CH_2CH_2COOH$ _____

2. Write the condensed structural formula for 2,3-dimethylpentane.

3. How many hydrogen atoms are there in 2,4-dibromo-2,3-dimethylhexane?

4. Identify the class of each of the following compounds:

    (a) CH_3CH_2OH _____

    (b) $CH_3CH_2CH \equiv CH_2$ _____

    (c) $HOOCCH_2CH_2COOH$ _____

    (d) $CH_3CH_2CH_2CHO$ _____

    (e) $CH_3OCH_2CH_2CH_3$ _____

5. Write the condensed structural formula for 3,4-dimethyloctane.

6. How many hydrogen atoms are there in 2,5-dichloro-3,4-dimethylheptane?

7. Identify the class of each of the following compounds:

   (a) $CH_3CH_2CH_2COOH$ _____

   (b) $CH_3CH_2OCH_2CH_3$ _____

   (c) $CHCCH_3$ _____

   (d) $CH_3CH_2CH_2CH_2CHO$ _____

8. Write the line formula for the simplest compound of the following classes:

   (a) Aldehyde _____

   (b) Ketone _____

   (c) Alkene _____

   (d) Acid _____

9. Write the formula for one isomer for each of the following compounds:

   (a) CH_3CH_2OH _____

   (b) $CH_3CH_2NH_2$ _____

   (c) $HCOOCH_3$ _____

   (d) CH_3COCH_3 _____

10. For each of the compounds you formed in question 9, give the class of compound to which the isomer belongs:

(a) _____

(b) _____

(c) _____

(d) _____

11. Write the line formula for the following compounds:

(a) 2,3-dimethylpentane   _____

(b) 2,2,4,4-tetramethylhexane   _____

12. Name the following compounds:

(a) $CH_3CH_2CH_2CH_2OH$ _____

(b) $CH_3CH_2CH_2OCH_2CH_2CH_3$ _____

(c) $CH_3CH_2CH_2CH_2CH_2CHO$ _____

(d) $CH_3CH_2CH_2CH_2COOH$ _____

13. Write the formula for the four isomers of C_4H_9Cl.

14. Name the compounds written in Problem 13.

   (a) _____

   (b) _____

   (c) _____

   (d) _____

15. What is the difference between a monomer and a polymer?

16. Draw a molecule of ethylene and show how the polymer polyethylene is formed.

Chapter 19: Experiment at Home

Grandma's Lye Soap

A soap is made by the base-catalyzed hydrolysis (saponification) of a triester such as a triglyceride found in animal fat. The ingredients for the manufacture of lye soap can be found in most households. Be extremely careful when performing this lab. There are hazards which are indicated.

Materials: Red Seal™ lye or a *solid* drain opener containing sodium hydroxide that does *not* contain aluminum. Drano™ will not work.
lard
vegetable oil
sodium chloride
porcelain or glass pot   Do not use aluminum.
glass measuring cup
wooden spoon
hot plate or stove
strainer
paper towels

WEAR SAFETY GOGGLES WHEN PERFORMING THIS LAB!!

Method: Place about $1/8$ cup of lard in the pot. In a glass measuring cup place $1/2$ cup of water and add to it three tablespoons of the lye. Stir carefully. The solution will become *very hot*. When the solution cools, add it carefully to the lard in the pot. Heat the lard and lye mixture until it boils. Turn down the heat and maintain an even boil with constant stirring until the mixture is almost solid. Remove from the heat and allow the soap to cool.

While the soap is cooling, mix two tablespoons of sodium chloride into $1/4$ cup of water. When the soap is cool enough to touch, add the salt solution to it and mix thoroughly. This will wash out any excess sodium hydroxide. Strain out the soap in the strainer and pat dry on paper towels.

You can make a different product by using $1/8$ cup of vegetable oil in place of the lard. Follow the same procedure and compare the resulting product to that obtained above.

Chapter 20: Nuclear Reactions

☐ Look over the introduction to this chapter. The comparison of nuclear and chemical reactions will help you understand some of the material in this chapter. Write these into your notes and remember the differences.

☐ Review Section 3.3 where you learned about subatomic particles before you continue with this chapter. Table 20.1 is the same table you saw in Chapter 3.

Section 20.1: Natural Radioactivity

☐ Read Section 20.1.

Most students find that working with nuclear equations is quite easy. The mass and atomic numbers must simply add to the same total on both sides of the equation. One confusing aspect of nuclear equations is that when beta particles are emitted, the atomic number is increased by one. The best way to work these problems is just to add the numbers on both sides and make sure they have the same total.

Disintegration series are identified by four classifications. The mass number divided by four plus any remainder gives the name of the series. Figure 20.2 gives three series. Figure 20.1 (a) shows the $4n + 2$ series.

Define the following terms as they relate to Section 20.1:

isotope: _____

alpha particle: _____

beta particle: _____

gamma particle: _____

radioactive decay: _____

disintegration: _____

event: _____

electromagnetic radiation: _____

parent isotope: _____

daughter isotope: _____

radioactive series: _____

Geiger counter: _____

tracers: _____

☐ Make sure to read the Examples and work the Practice Problems in the text as you read this section. Check your answers carefully.

☐ Work the Problems for this section at the end of the chapter.

Section 20.2: Half-Life

☐ Read Section 20.2.

It is easy to see that the fraction of the radioisotope that disintegrates is the same regardless of the amount of material present. In other words, if you start with 100 grams of sample, after one half-life 50% of the sample remains, or in this case, 50 grams. If you start with 10 grams of sample, after one half-life 50% of the sample remains, or in this case 5 grams. After two half-lives only 25% of the original material remains. The passage of three half-lives leaves only 12.5% of the starting material.

Using the fractional equivalents of the percentages, an equation can be developed. The fraction of material remaining after one half-life is:

$$\frac{N}{N_o} = \frac{1}{2^1} = \frac{1}{2} = 50\%$$

The fraction remaining after two half-lives is:

$$\frac{N}{N_o} = \frac{1}{2^2} = \frac{1}{4} = 25\%$$

The fraction remaining after three half-lives is:

$$\frac{N}{N_o} = \frac{1}{2^3} = \frac{1}{8} = 12.5\%$$

From this we can conclude that the fraction of starting radioisotope after n half-lives will be:

$$\frac{N}{N_o} = \frac{1}{2^n}$$

Once the fraction is calculated, it is multiplied by the radioisotope's half-life to determine the time required for the disintegration.

You can work Example 20.10 in this way.

$$\frac{N}{N_o} = \frac{1}{2^n}$$

$$\frac{4.92 \times 10^{22}}{3.76 \times 10^{23}} = \frac{1}{2^n}$$

$$\frac{3.76 \times 10^{23}}{4.92 \times 10^{22}} = \frac{2^n}{1}$$

$$7.64 = 2^n$$

Solving the exponential equation:

$$\log 7.64 = n \log 2 \qquad [\log 2 = 0.301]$$

$$\frac{\log 7.64}{\log 2} = n$$

$$n = 2.93 \text{ half-lives}$$

The time in years is simply the number of half-lives times the length of one half-life.

$$t = (2.93 \text{ half-lives})(3.31 \text{ years/half-life})$$

$$= 9.71 \text{ years}$$

Please notice that this method and the method in the text result in the same answer.

You can now follow the equation in the text for the rest of the examples.

Radioactive dating is a method of determining age. Carbon-14 is the isotope taken into living organisms in the form of CO_2. When death occurs, respiration ceases and no more carbon-14 is inhaled. At this time the isotope begins to decay without being replenished. If an approximate concentration is assumed for the organism at the time of death, then using the above equation the number of half-lives can be determined and hence the age of the specimen.

☐ Make sure to read the Examples and work the Practice Problems in the text as you read this section. Check your answers carefully.

☐ Work the Problems for this section at the end of the chapter.

Section 20.3: Nuclear Fission

Nuclear fission is the *splitting* of radioactive nuclei.

☐ Read Section 20.3.

Define the following terms in your own words:

atom smasher: _____

nuclear fission: _____

transmutation: _____

neutron: _____

proton: _____

deuteron: _____

positron: _____

chain reaction: _____

critical mass: _____

control rods: _____

☐ Make sure to read the Examples and work the Practice Problems in the text as you read this section. Check your answers carefully.

☐ Work the Problems for this section at the end of the chapter.

Section 20.4: Nuclear Fusion

Nuclear fusion is the *joining* of radioactive nuclei.

☐ Read Section 20.4.

☐ Make sure to read the Examples and work the Practice Problems in the text as you read this section. Check your answers carefully.

☐ Work the Problems for this section at the end of the chapter.

Chapter 20: Finishing Up

☐ Carefully read the summary section at the end of the chapter. Do you understand each paragraph? Do you know the terms used? If not, review the section indicated at the end of the paragraph.

☐ The section titled "Items for Special Attention" gives you some pointers to help you learn the material correctly and also can keep you from making some of the more common mistakes. Read these carefully.

☐ Work the "Self-Tutorial Problems." Check each answer as you complete the problem. If you have trouble with one of these, be sure to ask your instructor for help.

☐ Look over the list of "Key Terms" at the beginning of the chapter. If you do not recognize a term or are unsure of how it was used in the chapter, go back to that section and reread it.

☐ Use your lecture notes and the text to find the important topics covered in class. Make flash cards so that you can study these carefully before the exam.

☐ Work each of the problems at the end of the chapter in your text. Be sure to check your answers. If the answer is not given for that problem, work another that is similar and check the answer. If you are incorrect, check your work and review the chapter to see if you can answer the question correctly. If you cannot, get help from your instructor.

☐ Work the sample exam questions on the next few pages.

☐ Check your answers when you have completed all of them.

☐ Make up your own exam and exchange it with a fellow student or use it in your study group.

Chapter 20: Sample Exam Questions

1. A scientist prepared 78.0 grams of lanthanum-135. At the end of 58.5 hours, only 9.75 grams of the radioactive isotope remained. What is the half-life of lanthanum-135?

2. Complete the following equations:

   (a) $^{190}Pt \rightarrow \alpha +$ _____

   (b) $^{35}S \rightarrow \beta +$ _____

   (c) _____ $\rightarrow ^{140}Ba + \beta$

3. Alchemists dreamed of being able to turn base metals such as lead into gold. Mercury-198 can be converted into gold-198 by bombarding the mercury with neutrons. A proton is also produced. Write an equation for this process.

4. Iodine-131 is used to detect thyroid tumors. Iodine-131 has a half-life of 8.0 days. If a patient is given 0.008 µg of the radioisotope, how much will remain after 32 days?

5. Complete the following equations:

   (a) $^{208}\text{Fr} \rightarrow \alpha + $ _____

   (b) $^{90}\text{Sr} \rightarrow \beta + $ _____

   (c) _____ $\rightarrow {}^{242}\text{Pu} + \alpha$

6. Element 109 (^{266}Une) was prepared by bombardment of bismuth-209 atoms with iron-58. Write a balanced nuclear equation for this reaction.

7. Uranium-238 is an alpha emitter. Write a balanced nuclear equation showing this decay.

8. Phosphorus-32 is a beta emitter. Write a balanced nuclear equation showing this decay.

9. What was the starting material that produced the following products?

   (a) _____ $\rightarrow {}^{140}\text{Ba} + \beta$

   (b) _____ $\rightarrow {}^{234}\text{Th} + \alpha$

10. Complete the following equations and tell which of the following reactions is an example of fusion and which is an example of fission? (Hint: each line represents only one product.)

(a) $^{109}Ag + \alpha \rightarrow$ _____

(b) $^{10}B + \alpha \rightarrow$ _____

(c) $^{235}U + n \rightarrow {}^{160}Sm + {}^{72}Zn +$ ____ n

11. Phosphorus-32 is used in medical diagnosis to detect eye tumors. It has a half-life of 14.3 days. How much of an original 10.0 mg radioactive sample will be left after 71.5 days?

12. What was the original dose of Technetium-99m if after 2.0 days there is 0.39 gram left? The half-life of Technetium-99m is 6.0 hours.

13. Write a balanced nuclear equation for each of the following:

(a) the loss of an alpha particle by ^{218}Po

(b) the loss of a beta particle by ^{214}Bi

14. Calculate the number of grams of matter that need to be completely converted to energy in order to produce enough heat to warm 500 mL of water from 25°C to 60°C.

Answers to Sample Exam Questions

Chapter 1 -- Page 9
1. CO, H_2O, NO, NH_3
2. K, Mo, Mn, Bi
3. (a) C, (b) P, (c) P, (d) C, (e) P, (f) P
4. (a) iron, (b) gold, (c) mercury, (d) copper, (e) lead, (f) sodium, (g) silver, (h) phosphorus
5. (a) C, (b) C, (c) E, (d) E, (e) C, (f) C
6. (a) Fe, (b) Na, (c) Cl, (d) S, (e) K, (f) Au
7. See Figure 1.5 of your textbook.
8. (a) I, (b) E, (c) I, (d) I, (e) E
9. (a) P, (b) C, (c) P, (d) P, (e) C
10. (a) C, (b) C, (c) P, (d) P
11. Mass measures the amount of matter, weight takes into account the force of gravity on the mass.
12. (a) n, (b) m, (c) m, (d) m, (e) m
13. (a) mg, (b) mg, (c) t, (d) it, (e) mg
14. (a) am, (b) cm, (c) h, (d) ng, (e) aem
15. Groups are columns, periods are rows.

Chapter 2 -- Page 23
1. 13.3 pounds
2. 2.54×10^{-2} m
3. 4.54×10^5 g
4. (a) 3.5781×10^4, (b) 2.35×10^{-8}, (c) 2.365×10^5, (d) 9.872×10^{-9}, (e) 5.781×10^6, (f) 6.892×10^{-5}, (g) 1.4687×10^8, (h) 8.312×10^{-18}
5. 0.507 g/cm^3
6. yes
7. (a) 3.5×10^5, (b) 9.814 cm^3, (c) 6.75 mm, (d) 0.00213 L, (e) 27,000 µg
8. 2.25×10^{20} kg
9. (a) 8.78×10^{-24} cm^3, (b) 12.1 g/cm^3
10. 2.70 g/mL, aluminum
11. 2×10^2 square miles
12. 2×10^3 red blood cells

Chapter 3 -- Page 33
1. (a) ^{55}Mn, 25, 30, 25 (b) ^{18}O, 8, 8, 10 (c) ^{27}Al, 27, 13, 13 (d) ^{25}Mg, 12, 25, 12, (e) ^{20}Ne, 10, 10, 10 (f) ^{11}B, 5, 5, 6 (g) ^{54}Cr, 54, 24, 24 (h) ^{54}Fe, 26, 54, 26
2. 32.06 amu
3. 65.37 amu
4. 39.19% oxygen
5. (a) ^{209}Bi, 83, 126, 83 (b) ^{89}Y, 39, 39, 50 (c) ^{7}Li, 3, 7, 3 (d) ^{141}Pr, 59, 82, 59 (e) ^{64}Zn, 30, 64, 30
6. 28.1 amu
7. H_2S, KCl, SiH_4, CaO, $CdCl_2$
8. (a) atoms (b) molecules (c) whole-number (d) mass
9. For 1.0 g of H in each compound, the masses of C are 6.0 and 12.0 g. The ratio is 1:2.
10. electron, -1, 0, outside nucleus
    proton, +1, 1, nucleus
    neutron, 0, 1, nucleus

Chapter 4 -- Page 46
1. -3,-2,-1,0,+1,+2,+3
2. -1,0,+1
3. 0,1,2
4. 0,1
5. d
6. f
7. $1s^22s^22p^63s^2$
8. $1s^22s^22p^63s^23p^4$
9. $1s^22s^22p^3$
10. $1s^22s^22p^63s^23p^64s^2$
11. (a) Group VIA, (b) Group IIA, (c) Group VIIA, (d) Group VA
12. (b),(c),(d),(a)
13. [1,0,0,-1/2], [1,0,0,+1/2], [2,0,0,-1/2], [2,0,0,+1/2], [2,1,0,-1/2]
14. (a) $3d$, (b) $2p$ (c) $1s$ (d) $5f$ (e) $4f$
15. (a) 10 (b) 6 (c) 2 (d) 14 (e) 14
16. $1s^22s^22p^63s^23p^64s^23d^{10}4p^5$
17. (a) 4 (b) 3 (c) 2 (d) 1
18. $4s,4p,4d,4f$
19. 3
20. $1s^22s^22p^63s^23p^64s^2$
21. $[Rn]7s^26d^15f^3$
22. (a) $4d$ (b) $5p$ (c) $3d$ (d) $4d$ (e) $6p$
23. outermost shell electrons

Chapter 5 -- Page 57
1. Al^{3+} and SO_4^{2-} NH_4^+ and PO_4^{3-}
   Ca^{2+} and PO_4^{3-} NH_4^+ and SO_4^{2-}
2. :Ö–S̈=Ö:
3. :Ö–S–Ö:
      ‖
     :Ö:
4. [:Ö–Cl–Ö: with :Ö: above and :Ö: below]⁻

5. [Lewis structure of CO_3^{2-}]

6. Cl
7. 2 nitrogens, 8 hydrogens, 1 sulfur, and 4 oxygens
8. Most elements have eight electrons in their outer shells. Hydrogen has two electrons. For other exceptions, see text.
9. $1s^2 2s^2 2p^6 3s^2 3p^6$
10. (a) $MgCl_2$ (b) Al_2S_3 (c) Cs_2O (d) CaO (e) Li_3N
11. CO_2
12. Na^+ [Lewis structure of NO_3^-]
13. Ca^{2+} [Lewis structure of CO_3^{2-}]
14. (a) Na^+ and SO_4^{2-} (b) Ca^{2+} and CO_3^{2-} (c) K^+ and $C_2H_3O_2^-$ (d) K^+ and CN^- (e) Li^+ and OH^-
15. $Al(OH)_3$, $Al_2(SO_4)_3$, $AlPO_4$, $Ca(OH)_2$, $CaSO_4$, $Ca_3(PO_4)_2$, NH_4OH, $(NH_4)_2SO_4$, $(NH_4)_3PO_4$
16. F 17. T 18. F 19. F 20. T

Chapter 6 -- Page 67
1. (a) copper(II) nitrate, (b) nitrous acid, (c) phosphorus pentachloride, (d) nitrogen dioxide, (e) magnesium oxide
2. (a) Na_3PO_4, (b) FeF_2, (c) $Cr_2(SO_4)_3$, (d) H_2SO_3, (e) N_2O_4
3. (a) copper(I) chloride, (b) nitric acid, (c) phosphorus trichloride, (d) nitrogen monoxide, (e) manganese(II) oxide
4. (a) $Ca_3(PO_4)_2$, (b) FeF_3, (c) $MgSO_4$, (d) H_2SO_4, (e) NO_2
5. (a) Phosphorus pentachloride (b) dinitrogen pentoxide (c) iodine heptafluoride (d) bromine pentafluoride (e) chlorine dioxide
6. (a) potassium sulfate (b) ammonium carbonate (c) sodium hydroxide (d) magnesium phosphate (e) silver nitrate
7. (a) $MgCO_3$ (b) Na_2CrO_4 (c) $KMnO_4$ (d) $NH_4C_2H_3O_2$ (e) $Na_2Cr_2O_7$
8. (a) copper(II) hydroxide (b) silver sulfate (c) lead(IV) chloride (d) mercury(II) chloride (e) iron(III) oxide
9. (a) $Ni(C_2H_3O_2)_2$ (b) $AuCl_3$ (c) MnO_2 (d) $Sn(NO_3)_2$ (e) Cu_2S
10. (a) chloric acid (b) sulfuric acid (c) bromous acid (d) hypoiodous acid (e) phosphorous acid
11. (a) H_2SO_3 (b) HNO_3 (c) HIO_4 (d) $HBrO_2$ (e) H_3PO_2
12. (a) cadmium sulfate heptahydrate (b) cobalt(II) fluoride (c) tin(IV) chromate (d) selenium trioxide (e) hypochlorous acid
13. (a) $ZnCrO_4$ (b) $Mg(MnO_4)_2$ (c) $Pb(C_2H_3O_2)_4$ (d) HCl (e) $NaCl$

Chapter 7 -- page 79
1. C_3H_6O
2. $C_3H_6O_2$
3. 2.56×10^{23} hydrogen atoms
4. 3.97×10^{23} hydrogen atoms
5. (a) 77.0 amu (b) 759.6 amu
6. 30.4% nitrogen
7. (a) 0.283 mol (b) 0.438 mol
8. 3.94×10^{22} oxygen atoms
9. 464 grams
10. 3.82×10^{-23} grams
11. 2.65×10^{-22} grams
12. $C_5H_{11}ON$
13. C_2H_6O
14. C_6H_{12}

Chapter 8 -- Page 91
1. $2\ C_6H_{14} + 19\ O_2 \rightarrow 12\ CO_2 + 14\ H_2O$
2. $C_7H_{16} + 11\ O_2 \rightarrow 7\ CO_2 + 8\ H_2O$
3. $3\ NaOH(aq) + H_3PO_4(aq) \rightarrow Na_3PO_4(aq) + 3\ H_2O(l)$
4. $3\ Ba(OH)_2(aq) + 2\ H_3PO_4(aq) \rightarrow Ba_3(PO_4)_2(s) + 6\ H_2O(l)$
5. substitution $Zn(NO_3)_2 + Cu(s)$
6. combination MgO
7. (a) $2\ Al + 2\ NaOH + 2\ H_2O \rightarrow 2\ NaAlO_2 + 3\ H_2$
   (b) $3\ C + Fe_2O_3 \rightarrow 3\ CO + 2\ Fe$
8. (a) $S + O_2 \rightarrow SO_2$
   (b) $Zn + CuCl_2 \rightarrow ZnCl_2 + Cu$
   (c) $SnCl_2 + Cl_2 \rightarrow SnCl_4$
9. (a) $(NH_4)_2CO_3 + Ba(NO_3)_2 \rightarrow BaCO_3 + 2\ NH_4NO_3$
   (b) $2\ KClO_3 \rightarrow 2\ KCl + 3\ O_2$
   (c) $2\ Ag + S \rightarrow Ag_2S$

10. (a) HCl + NaOH → NaCl + H$_2$O
(b) Na$_2$CO$_3$ + 2 HCl →
2 NaCl + CO$_2$ + H$_2$O
(c) Pb(NO$_3$)$_2$ + 2 NaCl →
PbCl$_2$ + 2 NaNO$_3$
11. (a) 2 H$_2$ + O$_2$ → 2 H$_2$O
(b) Mg + 2 HCl → MgCl$_2$ + H$_2$
(c) Na$_2$S + Ni(NO$_3$)$_2$ →
NiS + 2 NaNO$_3$
12. (a) 2 C$_6$H$_{14}$ + 19 O$_2$ → 12 CO$_2$ + 14 H$_2$O
combustion reaction, carbon dioxide, and water
(b) Ba(OH)$_2$ + H$_2$SO$_4$ → BaSO$_4$ + 2 H$_2$O
double substitution, barium sulfate, and water
(c) SO$_3$ + CaO → CaSO$_4$
combination, calcium sulfate
13. (a) NH$_4$HCO$_3$ → NH$_3$ + CO$_2$ + H$_2$O
decomposition, ammonia, carbon dioxide, and water
(b) 2 AlCl$_3$ + 3 Na$_2$CO$_3$ →
Al$_2$(CO$_3$)$_3$ + 6 NaCl
double substitution, aluminum carbonate, and sodium chloride
(c) 4 Al + 3 O$_2$ → 2 Al$_2$O$_3$
combination, aluminum oxide
14. (a) NR, (b) Ba$_3$(PO$_4$)$_2$, (c) Al$_2$(SO$_4$)$_3$, (d) NR

Chapter 9 -- Page 101
1. 2 H$^+$(aq) + CO$_3^{2-}$(aq) → CO$_2$(g) + H$_2$O(l)
2. Ba^{2+}(aq) + 2 OH$^-$ + 2 H$^+$ + SO$_4^{2-}$(aq) →
BaSO$_4$(s) + 2 H$_2$O(l)
3. H$^+$(aq) + HCO$_3^{2-}$(aq) → CO$_2$(g) + H$_2$O(l)
4. (a) no (b) yes (c) no (d) yes (e) no
5. (a) 3 CaCl$_2$ + 2 K$_3$PO$_4$ →
Ca$_3$(PO$_4$)$_2$ + 6 KCl
3 Ca^{2+}(aq) + 2 PO$_4^{3-}$(aq) → Ca$_3$(PO$_4$)$_2$(s)
(b) FeCl$_2$ + 2 NaOH → Fe(OH)$_2$ + 2 NaCl
Fe^{2+}(aq) + 2 OH$^-$(aq) → Fe(OH)$_2$(s)
(c) 2 NaHCO$_3$ + H$_2$SO$_4$ →
2 CO$_2$ + 2 H$_2$O + Na$_2$SO$_4$
H$^+$(aq) + HCO$_3^{2-}$(aq) → CO$_2$(g) + H$_2$O(l)
(d) no net ionic reaction
(e) Ba(OH)$_2$ + 2 HCl → BaCl$_2$ + 2 H$_2$O(l)
H$^+$ + OH$^-$ → H$_2$O(l)
6. Cl$^-$
7. H$_2$SO$_4$ + 2 KOH → K$_2$SO$_4$ + 2 H$_2$O
H$^+$ + OH$^-$ → H$_2$O
8. Fe(NO$_3$)$_3$ + 3 NaOH →
Fe(OH)$_3$ + 3 NaNO$_3$
Fe^{3+} + 3 OH$^-$ → Fe(OH)$_3$(s)
9. Zn + 2 HCl → ZnCl$_2$ + H$_2$

Zn + 2 H$^+$ → Zn^{2+} + H$_2$
10. NaHCO$_3$ + HCl →
CO$_2$ + H$_2$O + NaCl
H$^+$(aq) + HCO$_3^-$(aq) → CO$_2$(g) + H$_2$O(l)
11. AgNO$_3$ + NaCl → AgCl(s) + NaNO$_3$
Ag$^+$ + Cl$^-$ → AgCl(s)

Chapter 10 -- Page 110
1. 105.5 g AgCl
2. no, 6.08 g KCl when complete
3. 92.9 g BaCO$_3$
4. no, 16.58 g MgO
5. 0.764 g H$_2$
6. 221.9 g O$_2$
7. 492 g NaAlO$_2$
8. 9.20 x 10^{24} oxygen atoms
9. 6.48 L C$_2$H$_6$O
10. 24.6 g CO$_2$
11. 78.0%
12. 6 x 10^{13} molecules
13. C$_6$H$_7$N
14. 150 kg NaOH
15. 0.0426 g K$_2$Cr$_2$O$_7$
16. 0.01654 g Na$_2$SO$_4$

Chapter 11 -- Page 119
1. [Na$^+$] = 0.186 M, [Mg^{2+}] = 0.280 M, [NO$_3^-$] = 0.744 M
2. 84.5 mL
3. 0.1884 M HCl
4. [Na$^+$] = 0.0324 M, [Mg^{2+}] = 0.350 M, [Cl$^-$] = 0.730 M
5. 32.8 mL
6. 0.136 M
7. 0.156 M
8. 41.9 g CaCl$_2$
9. 99.6 mL
10. 0.04113 M
11. [NH$_4^+$] = 4.35 M
[PO$_4^{3-}$] = 1.45 M
12. [Na$^+$] = 0.00308 M
[Al^{3+}] = 0.0114 M
[Cl$^-$] = 0.0373 M
13. 0.2785 M
14. 67.6 mL
15. 0.01985 mol BaSO$_4$ precipitates
[Ba^{2+}] = 0.2511 M
[Cl$^-$] = 0.8892 M
[Na$^+$] = 0.3870 M
16. 45.3 mL

Chapter 12 -- Page 131
1. Gas 1 $V_2 = 406$ mL
   Gas 2 $P_2 = 1.48$ atm
   Gas 3 $T_1 = 104$ K
   Gas 4 $P_1 = 6.97$ atm
2. 2.60 L O_2
3. 956°C
4. 85.5 L
5. 13.3 L
6. 3.12 atm
7. 1.93 atm
8. 69.4 mL
9. 1040 L
10. 2.9 hours
11. 8.29 L
12. 118 g CO_2
13. 2.05 g/L
14. 0.406 g O_2
15. $C_4H_6O_2$

Chapter 13 -- Page 143
1. Mg
2. Be
3. S
4. Ba
5. Na, Cl^-, Mg, O^{2-}
6. S
7. Ba
8. C-H(0.4), C-N(0.5), Si-O(1.7), K-O(2.7), Li-F(3.0)
9. bent, triangular, linear, bent, trigonal pyramidal
10. polar, nonpolar, nonpolar, polar, polar
11. CO

Chapter 14 -- Page 151
1. 110 g
2. 6870 J
3. 52.2 g
4. 3.05 kJ/g
5. 1.64 kJ
6. 5.74×10^7 J
7. (a) fusion, (b) sublimation, (c) vaporization
8. must overcome all intermolecular forces

Chapter 15 -- Page 159
1. (a) W (b) G (c) W
2. 1.24 M
3. 73.2 g/mol
4. 1.679 M
5. 11 atm
6. Polarity of N-H bond, H bonding
7. 0.03333 m
8. 11.2 mol
9. $X_{H_2O} = 0.684$, $X_{CH_3CH_2OH} = 0.316$
10. 25.6 m CH_3CH_2OH
11. 1.88 m
12. $C_{10}H_{14}N_2$
13. 5.1°C/m

Chapter 16 -- Page 171
1. $2 H^+ + 2 MnO_4^- + 3 H_2S \rightarrow$
   $2 MnO_2 + 3 S + 4 H_2O$
2. 33.5 meq
3. 0.00218 eq
4. $S_2O_3^{2-} + 4 Cl_2 + 5 H_2O \rightarrow$
   $2 SO_4^{2-} + 8 Cl^- + 10 H^+$
5. 0.15 eq/L = 0.15 N
6. 0.06962 N
7. $3 As_2O_3 + 4 CrO_4^{2-} + 20 H^+ \rightarrow$
   $6 H_3AsO_4 + 4 Cr^{3+} + H_2O$
8. $Pb + PbO_2 + 2 H_2SO_4 \rightarrow 2 PbSO_4 + 2 H_2O$
9. $H_2Cr_2O_7 + 3 Zn + 12 H^+ \rightarrow$
   $2 Cr^{3+} + 3 Zn^{2+} + 7 H_2O$
10. (a) chlorine(IV) oxide
    (b) chlorine(V) oxide
    (c) chlorine(VI) oxide
    (d) chlorine(VII) oxide
11. (a) $SnCl_2$ and $SnCl_4$
    (b) $SeBr_2$ $SeBr_4$ and $SeBr_6$
12. 0.08755 N
13. $KMnO_4$, O_2, $K_2Cr_2O_7$
14. Zn is oxidized, Cu^{2+} is reduced, Cu^{2+} is the oxidizing agent, Zn is the reducing agent
15. P_4 is oxidized, P_4 is reduced, P_4 is the oxidizing agent, P_4 is the reducing agent
16. 4581 g/eq

Chapter 17 -- Page 181
1. (a)
$$\frac{[O_2]^7[NH_3]^4}{[NO_2]^4[H_2O]^6}$$
(b) R, R, L, L, no effect
2. (a)
$$\frac{[HF]^3 [PF_3]}{[F_2]^3 [PH_3]}$$
(b) no effect, R, R, L, no effect
3. (a) $COCl_2 \rightleftharpoons CO + Cl_2$
   (b) $2 HBr \rightleftharpoons H_2 + Br_2$
   (c) $6 CO_2 + 6 H_2O \rightleftharpoons 9 O_2 + 2 C_3H_6$
4. 1.51
5. 378°C

6.
$$\frac{[C_2H_4Cl_2][HCl]^2}{[C_2H_6][Cl_2]^2}$$

7. (a) R (b) L (c) R
8. (a) R (b) R (c) L (d) R
9. 50
10. 1.5 M
11. 1.1 M
12. 0.002 M/s

Chapter 18 -- Page 193
1. (a) 10.839 (b) 9.731
2. (a) 1.7×10^{-9} (b) 7.6×10^{-1}
   (c) 1.08×10^{-6}
3. 9.56
4. (a) 7.570 (b) 2.978
5. (a) 4.6×10^{-11} (b) 7.6×10^{-2}
   (c) 1.08×10^{-3}
6. 2.41
7. (a) SO_4^{2-} (b) HPO_4^{2-} (c) CO_3^{2-} (d) OH^-
8. (a) H_3O^+ (b) H_3PO_4 (c) H_2SO_4
   (d) H_2CO_3
9. (a) 9.629 (b) 4.277 (c) 7.047 (d) 6.502
10. (a) 2.10×10^{-2} (b) 1.04×10^{-7}
    (c) 3.52×10^{-12} (d) 1.65×10^{1}
11. 1.90×10^{-3} M
12. 4.69%
13. (a) 4.371 (b) 9.723 (c) 6.953 (d) 7.498
14. 3.19×10^{-11}
15. (b) and (d)
16. (a) basic (b) basic (c) acidic
    (d) acidic (e) neutral

Chapter 19 -- Page 208
1. (a) alcohol (b) aldehyde (c) ether
   (d) alkene (e) carboxylic acid
2. $(CH_3)_2CHCH(CH_3)CH_2CH_3$
3. 16
4. (a) alcohol (b) alkyne (c) carboxylic acid
   (d) aldehyde (e) ether
5. $CH_3CH_2CH(CH_3)CH(CH_3)(CH_2)_3CH_3$
6. 18
7. (a) carboxylic acid (b) ether (c) alkyne
   (d) aldehyde
8. (a) HCHO (b) CH_3COCH_3 (c) CH_2CH_2
   (d) HCOOH
9. (a) CH_3OCH_3 (b) CH_3NHCH_3
   (c) CH_3COOH (d) CH_3CH_2CHO
10. (a) ether (b) amine (c) acid (d) aldehyde
11. (a) $(CH_3)_2CHCH(CH_3)CH_2CH_3$
    (b) $(CH_3)_3CCH_2C(CH_3)_2CH_2CH_3$
12. (a) butanol (b) dipropyl ether (c) hexanal
    (d) pentanoic acid
13. $CH_3CH_2CH_2CH_2Cl$ $CH_3CH_2CHClCH_3$
    $(CH_3)_2CHCH_2Cl$ $(CH_3)_3CCl$
14. 1-chlorobutane, 2-chlorobutane
    1-chloro-2-methylpropane,
    2-chloro-2-methylpropane
15. monomer is single unit, polymer is a combination of many monomers
16. see Section 19.4 of your textbook

Chapter 20 -- Page 219
1. 19.5 hours
2. (a) ^{186}Os (b) ^{35}Cl (c) ^{140}Cs
3. $^{198}Hg + ^{1}n \rightarrow ^{198}Au + ^{1}p$
4. 0.0005 μg
5. (a) ^{204}At (b) ^{90}Y (c) ^{246}Cm
6. $^{209}Bi + ^{58}Fe \rightarrow ^{266}Une + ^{1}n$
7. $^{238}U \rightarrow \alpha + ^{234}Th$
8. $^{32}P \rightarrow \beta + ^{32}S$
9. (a) $^{140}Cs \rightarrow ^{140}Ba + \beta$
   (b) $^{238}U \rightarrow ^{234}Th + \alpha$
10. (a) $^{109}Ag + \alpha \rightarrow ^{113}In$ (fusion)
    (b) $^{10}B + \alpha \rightarrow ^{14}N$ (fusion)
    (c) $^{235}U + ^{1}n \rightarrow ^{160}Sm + ^{72}Zn + 4\ ^{1}n$ (fission)
11. 0.313 mg
12. 100 g
13. (a) $^{218}Po \rightarrow \alpha + ^{214}Pb$
    (b) $^{214}Bi \rightarrow \beta + ^{214}Po$
14. 8.1×10^{-13} kg